SILFS

Volume 2

Open Problems in Philosophy of Sciences

Volume 1
New Essays in Logic and Philosophy of Science
Marcello D'Agostino, Giulio Giorello, Federico Laudisa, Telmo Pievani,
and Corrado Sinigaglia, eds.

Volume 2
Open Problems in Philosophy of Sciences
Pierluigi Graziani, Luca Guzzardi, and Massimo Sangoi, eds.

SILFS Series Editor
Marcello D'Agostino   marcello.dagostino@unife.it

Open Problems in Philosophy of Sciences

edited by

Pierluigi Graziani
Luca Guzzardi
and
Massimo Sangoi

ISBN 978-1-84890-062-2

College Publications
Scientific Director: Dov Gabbay
Managing Director: Jane Spurr
Department of Computer Science
King's College London, Strand, London WC2R 2LS, UK

http://www.collegepublications.co.uk

Original cover design by orchid creative   www.orchidcreative.co.uk

Printed by Lightning Source, Milton Keynes, UK

Reflecting on Open Problems in Philosophy of Sciences

Pierluigi Graziani
University of Urbino
pierluigi.graziani@uniurb.it

Luca Guzzardi
Dipartimento di Fisica, University of Pavia
INAF Osservatorio Astronomico di Brera, Milan
luca.guzzardi@unipv.it

Massimo Sangoi
University of Urbino
massimosangoi@gmail.com

There are two unfortunately widespread positions which concern the relationship between philosophy and science:

The first rises from the fear of some philosophers that the interaction between philosophy and science could result in an expansion of the latter and a reduction of the former; such a fear favors philosophical efforts for the construction of rigid boundaries between science and philosophy;

The second position, although encourages philosophy to analyze scientific issues and take them as starting points for philosophical reflections, is however incapable of understanding the depth of those scientific issues.

The first position transforms philosophical analysis into a mere a priori activity, that is a sort of scholasticism; the second one, on the contrary, considers science, but only in an extrinsic way.

Fortunately, beyond these two positions a third one is possible, which consists, in general, in thinking of philosophy as the construction of a justified solution to a particular class of fundamental problems and of the philosopher as someone who can argue for his/her answers to the problems, who knows how to place them historically, who can rationally justify them and make them available for discussion and modification; and in particular, it consists in thinking that, when philosophy picks problems from specific sciences, it must analyze them with scientific competence, care, skill, by grasping their essence and putting them in a general perspective.

With these purposes, following (*mutatis mutandis*) the experience of two schools in Francavilla al Mare organized by the Italian Philosophical Society[1], the school *Open Problems in Philosophy of Sciences*[2] (Cesena, Italy, April 15–16–17, 2010) has promoted and stimulated a methodologically conscious and mature philosophical analysis, capable of dealing with precision and deep cultural awareness with the problems raised by some particular sciences.

The school, opened by such a distinguished professor of philosophy of science as Evandro Agazzi, has allowed and stimulated a dialogue among scholars from different Italian Universities and fueled a dialogue open to students in the field and teachers of every grade.

Rather than talking about the philosophy of science, scholars have shown how this discipline can and should be done by reflecting on important philosophical issues in the domains of life sciences (Giovanni Boniolo, Cecilia Nardini; Fridolin Gross; Fabio Lelli; Elena Casetta), mathematics (Mario Piazza, Gabriele Pulcini; Gianluca Ustori; Andrea Sereni; Valerio Giardino), psychology (Alfredo Paternoster, Maria-Erica Cosentino; Barbara Giolito; Maria Grazia Rossi; Maria Francesca Palermo), and physics (Vincenzo Fano, Giovanni Macchia; Claudio Mazzola; Giulia Giannini; Giuliano Torrengo).

---

[1] See: C. Tatasciore, P. Graziani, G. Grimaldi [eds.] *Prospettive Filosofiche 2006: Il Realismo*. Naples, Istituto Italiano per gli Studi Filosofici, 2007; C. Tatasciore, P. Graziani, G. Grimaldi [eds] *Prospettive Filosofiche 2009: Ontologia,* Naples, Istituto Italiano per gli Studi Filosofici, 2011.

[2] The school, organized by Mario Alai, Vincenzo Fano, Pierluigi Graziani and Gino Tarozzi was made possible by a synergy between the *Interuniversity Centre for Research in Philosophy and Foundations of Physics*, the *Italian Society of Logic and Philosophy of Science,* the *Department of Philosophy of University of Urbino*, and the *Municipality of Cesena*, which has been favoring for many years, mainly thanks to Franco Pollini, the Interuniversity Center in organizing major international events in the philosophy of sciences. See: http://synergia.jimdo.com/

Therefore, the importance of making those assets available has produced this book that intends to stimulate a correct philosophical analysis of various problems raised by some contemporary sciences.

The volume provides comprehensive and accessible coverage (including, an analysis of the major positions and battle lines) of some open problems of the disciplines of philosophy of the life sciences, philosophy of mathematics, philosophy of mind, and philosophy of physics. Each chapter breaks new ground: they not only present open problems, but advance possible solutions as well. In light of the depth and evolution of the disciplines today, no single volume can provide extensive coverage of all open problems, but most of the chapters of this book contain an extensive bibliography, and in total it provides a clear picture of the state of the art of some open problems in philosophy of sciences.

There are some overlaps between the contributed papers; this was explicitly encouraged and is to be expected: in real science the main issues and views of either discipline frequently permeate those of the other.

The book has four sections and each section has an introduction. After each introduction to the specific disciplines (philosophy of the life sciences, philosophy of mathematics, philosophy of psychology, and philosophy of physics) by Giovanni Boniolo, Mario Piazza, Alfredo Paternoster and Vincenzo Fano respectively, each section follows with four contributed papers on the open problems in that context.

Throughout the process of assembling this book, we benefited from the sage advice of colleagues and friends. Thanks especially to Mario Alai, Claudio Calosi, Marcello D'Agostino, Mauro Dorato, Vincenzo Fano, Corrado Sinigaglia, Gino Tarozzi and Isabella Tassani.

Table of contents

First Section
Philosophy of Biology

Introduction

Giovanni Boniolo
IEO (Istituto Europeo di Oncologia), Milan
Scuola Europea di Medicina Molecolare, Milan
Dipartimento di Medicina, Chirurga e Odontoiatria
University of Milan
giovanni.boniolo@ifom-ieo-campus.it

Over these last years, biology, biomedicine and medicine are making a great leap forward and the humanistic reflection on them cannot fall behind. This is a strong claim, but it wants to emphasize two aspects: 1) the amazing improvement of knowledge we are having with the molecular turn and thanks to the technological advancement; 2) the necessity for the humanistic analysis, be it foundational, historical, ethical or sociological, to keep up the pace and, thus, to prepare new categories by means of which to cope with the situation.

By taking what above for granted, we need a new generation of scholars with both humanistic and scientific competences. We need humanistic scholars able to read a scientific paper concerning the molecular bases of evolution or the post-genomic approach to a given pathology, capable of understanding the computational biologists' formalism or what a synthetic biologist is doing. Now, even more then in the past, it is no longer the time to do humanistic analysis of pieces of science without knowing those pieces. The uselessness of armchair philosophers of science, or ethicists, or sociologists lies on the fact that contemporary biology, biomedicine, and medicine touch the depth of the human "nature" (whatever the meaning of "nature" is) as never before. And touching it without the necessary knowledge on the status of art can produce dangerous outcomes, especially from the ethical and sociological point of view.

More or less, all the authors of the four papers of this section seem to be perfectly aware of this situation and try to move their analyses from an understanding of what real science is. In particular, Casetta espouses a totally sharable ontological conventionalism on the temporal boundaries of species; Gross argues for a non mechanistic, but systemic, explanation in molecular biology; Lelli, convincingly, advances the thesis of the impossibility of doing bioethics without a proper knowledge in the philosophy of the life sciences; and Nardini analyses the relevant topic of the *in silico* models in biology.

The Role of *in silico* Models in Evolutionary Studies

Cecilia Nardini
IFOM-IEO Campus and University of Milan
cecilia.nardini@ifom-ieo-campus.it

1 Introduction

The increasing speed and performance of computers has in recent years created the possibility to simulate real-world processes or phenomena involving a large number of individual entities, thereby opening the door to a new wave of research on complex systems. This trend can be seen, for instance, in geology and climate studies. Evolutionary biology is also moving in this direction, as simulations based on digital organisms are being developed for the in-depth exploration of the underlying principles of natural evolution. Evolutionary biology, through population genetics, is not new to numerical modelling; however, traditional mathematical models are built upon a simplification or an approximation of the (known) equations that govern the system such as the equations for the population dynamics, or the flow of energy through the trophic levels. In this context, the numerical model plays a purely computational role, providing an approximate numerical solution to the problem. The case of the so-called evolutionary algorithms (EA) like TIERRA (Ray 1999) or AVIDA (Lenski et al., 2003) is altogether different: the computer does not solve numerically the equations referred to a biological system, since the laws governing its dynamics are largely unknown. Rather, it provides the substrate on which a fully artificial system, which nonetheless is taken to be similar to the natural one in relevant respect, can develop.

Many evolutionary biologists see these computer models as true experiments that can elucidate, by observing their progression, some of the causal mechanisms underlying natural evolution (Lenski et al., 2003). The possibility to control the conditions and to repeat the trials at will makes simulations invaluable tools in the investigation of an issue that is intrinsically impervious to experimental approach. But to what extent, if at all, is this identification of simulation with experiment warranted? In what follows I will scrutinize the use of computer models in evolutionary studies, exploring the nature of the knowledge they produce, the reliability of their use, and the relation it bears to the understanding of real-world biological systems.

2 Evolutionary Algorithms and AVIDA

The term "simulation" in evolutionary studies refers to a collection of entities that "live" and "reproduce" inside the virtual space of the computer memory. These entities compete for some essential resource, typically memory space or CPU time, and their replication process is imperfect, yielding random mutations that may affect the entities' effectiveness in the competition. It is generally assumed that these features, together with the details of the interaction among the entities and between the entities and the virtual environment, is enough to make this system subject to an evolutionary dynamics akin to the one proper of biological systems. Simulating an evolutionary system *in silico* has the aim of discovering underlying mechanisms common to all evolving systems: in this sense the simulation is a model that stands for a whole class of systems, characterized by evolution as a dynamical process. In other words, digital organisms are not conceived to be models of biological organisms undergoing natural evolution but rather as separate entities instantiating the same evolutionary process on a different substrate. Indeed, when building those algorithms, the investigators often avoid altogether to try to model chemical or physical features of the system in order to capture other features they deem relevant, such as the topological structure of the interaction network (Taylor and Hallam 1997). The underlying implication is that the aim of the simulation is to be an instantiation of an abstract mathematical regularity rather than a more or less faithful reproduction of evolution as it happens in nature.

One important aspect of computer models resides in the possibility of freely manipulating them to an extent that is in most cases superior to what can be done on the corresponding physical system. This makes them a tool of great potential for the study of evolutionary biology, since in this field experiments with the biological components present difficulties of duration and complexity almost impossible to overcome. The so-called Evolutionary Algorithms (EA) have been created precisely for this purpose.

The computer program TIERRA was developed in the early 1990s by ecologist T.S. Ray (1991). He drew inspiration from *core wars*, a competition between programmers in which human-designed, self-replicating software compete for the computational resources needed for replication. Likewise, in TIERRA, the starting point is a collection of programs able to self-replicate that compete for resources; however, the replication mechanism is faulty, introducing random mutations in the code in the form of deletion, duplication or swapping of code lines. As time proceeds, the initial population branches into variants that differ in replication efficiency, setting in a dynamics of (natural) selection. Apart from the first replicator, which is designed, the digital "ecosystem" is out of the hands of its creator as it develops and evolves without further human interventions. Due to the random elements introduced by mutations and flawed replication, every run of the program results in a different evolutionary story. On the other hand, however, the evolution of the system depends totally on the set of rules that the programmer has originally decided, dictating how the digital organisms interact with each other and what is the form and strength of the competition between them. The program's rules represent the only constraint posed upon the behaviour of the components and the program follows them in a deterministic way.

As its precursor TIERRA, AVIDA is based on self-replicating competing programs. AVIDA was created for investigating the issue of irreducible complexity: Its objective was to show that complex functions can appear in evolving organisms building on simpler, existing features, without the need for a plan or design. To tackle with this issue the original program had to be modified and organisms had to be endowed with an additional trait independent from replication speed. In fact, in TIERRA it was found that replication speed is subject to a selective pressure that is too strong for complex features to develop (see Taylor and Hallam, 1997). Researchers decided to look at the programs' ability to perform simple logical operations, and in particular at the way such capacity could be built upon simpler instructions. In addition to the replication loop, the digital organisms in AVIDA include therefore a computational "metabolism" which allows them to obtain resources in reward for performing logical operations on input bit strings. The ancestor organism can replicate but cannot perform any logical function; as the algorithms mutate and evolve, these logical functions can be built up as appropriate combination of the existing instructions. As already mentioned, a program's ability to perform an operation is rewarded in the form of extra processing time. The important feature to be underlined is that the reward is increasing exponentially with the complexity of the operation: the most complex operations are also fragile (cf. Lenski et al., 2003) and for this reason it is necessary to boost the evolutionary advantage of the organisms possessing them.

3 The Epistemological Status of Computer Experiments

A thing that is striking upon reading Lenski et al. (2003) original paper about AVIDA is the pervasiveness of experiment talk that permeates it. The authors refer to the technical details of the runs as "experimental conditions" and the digital organisms are described using a biological wording – "genome", "phenotype", "lineage"–, that conveys the idea that AVIDA is indeed the experiment in evolution that biologists have ever been yearning for. It is quite clear, though, that EAs are ultimately not meant as proper simplification of biological organisms: the resemblance between the two is purely formal. The rudimental metabolism and reproductive mechanism of digital organisms stand just for *properties* of living organisms that the model incorporates because they are believed to be crucial for the process under investigation. The aim of EAs is rather to reproduce a particular causal link by putting it at work in isolation, within a tractable and controlled system that allows manipulability and repeatability.

By this token simulations bear a certain resemblance to experiments with model organisms; however, where model organisms are the result of a tailoring of a natural system, the computer model is completely artificial and does not contain anything else than what has been put into it. Resorting to our case study may make this distinction clearer. As discussed in Sec.1, it is the set of rules defined by the programming scientist that deterministically dictates, aside from the mutations introduced randomly, the steps the program will follow. Any robust regularity appearing in the results of different runs must therefore have its grounding in this set of rules. In experiments with model organisms, the presence of intervening factors other than the one under study can never be totally excluded, since the model organism is the result of an operation of purification or amplification of one within a collection of natural features. The computer model on the contrary is built from scratch and it incorporates from the beginning only the factors which are deemed relevant by the programming scientist: the behaviour it produces can therefore be ascribed with confidence to those factors alone. The above discussion is meant to point out that there's a suggestive way in which simulations can be thought of as experiments, but that this point of view has not to be taken too far, since simulations are permeated with theory more than any experimental setup.

The hybrid nature of simulations, their standing halfways between a theoretical model and an experimental setup, has been acknowledged in a mole of recent philosophical work: see for instance Hartmann (1996), Bedau (1999), Winsberg (1999), Fox Keller (2003), Peck (2004) and Lenhard (2007). Notably, E. Fox Keller has analysed the epistemological standing of the so-called Cellular Automata (CA) in a paper which has had considerable influence (Fox Keller, 2003): even though CA have important differences with respect to EAs, Keller's analysis significantly captures the nature of computer simulation as "trial theory" which is at the same time embodied by the program and put to test by it. In order to have the simulation include the key causal mechanisms at work, it is necessary for the scientist to make a hypothesis over the articulation of the phenomenon under investigation, and it is precisely this hypothesis that the simulation is embodying. Scientists observe the behaviour of evolutionary software under varying conditions or a change in the parameters, in the same way they would observe an experimental system respond to external stimuli. However, what they are looking for is not, like in physical experimental systems, a key to the explanation of an observed behaviour, since this behaviour ultimately depends only on the rules that have been knowingly set from the beginning. Rather, they are verifying precisely the capability of the computer system to produce the behaviour which corresponds to a physical phenomenon, because this capability would be a sign that the hypothesis underlying the computer model is adequate in accounting for the natural phenomenon. However, it is important to notice that the proof simulations can provide can in general only be a negative one. For instance, in the case of evolutionary algorithms, the experience with TIERRA showed that competition for memory and time tends to drive the programs towards an "optimized", short and simple, phenotype: this is a evidence that selective pressure alone is not sufficient to grant the conditions for the evolution of complexity (Ofria and Adami, 2002).

Drawing *positive* conclusions from simulations, on the other hand, is inherently much more problematic for a number of reasons that strictly depend on their peculiar epistemological status, which has been outlined above. Standard experiments suffer notoriously the problem of validation of the experimental setup and particularly the issue of artefacts, a problem that has been treated by A. Franklin (1989) and reprised, in the context of computer experiments, by E. Winsberg (1999). The fruitful use of an experimental setup in scientific enquiry closely depends on the scientist's possibility to distinguish artefacts produced by the apparatus from valuable results that are direct consequences of the natural phenomenon under investigation. In the case of computer experiments the issue is particularly intricate since, as we have seen, the simulation system is completely artificial and all the results it produces are, incontestably, the results of the rules set up by the human programmer.

Validation of an experimental system is usually achieved by checking if the experimental system is able to reproduce the real world behaviour that it is meant to model. However, in the case of EAs, there is not an accepted mechanism for the issue under investigation: therefore, what scientists do is to observe the evolution of digital organisms under different conditions and establish which – if any – hypothesis over the interaction is best reproducing the behaviour of the natural system. Reproducing the known phenomenon is then intended as a test directly on the theoretical presuppositions rather than as a check on the reliability of the experimental system. In other words, in evolutionary algorithms, the exploration and testing part come to be intertwined, as an effect of the hybrid nature of simulations. This has of course an influence on the interpretation of simulations and the limits of their use in scientific practice; in what follows I will argue that in the case of evolutionary algorithms the issue is strongly in place.

4 AVIDA: results and discussions

As mentioned above AVIDA was created with the purpose of investigating the issue of the origin of complexity by looking at the emergence, in a population of evolving computer programs, of the capacity to perform logical operations. The EQU function or identity evaluation was taken as the benchmark for the evaluation of complexity, since it is the operation requiring the largest assembly of available instructions (cf. Lenski et al., 2003). The researchers looked at the development of EQU function in the digital organisms, assessing whether and at what time point in evolution it arose. Furthermore they investigated its relation to other, simpler functions in order to detect possible fixed patterns or obligatory steps in evolution of the complex feature EQU. They found that EQU evolved in almost half of the cases (23 runs out of 50) and, by analyzing in detail one of the successful programs, they highlighted some features which seem to nicely resonate with natural evolution of biological systems. In fact, they found that oft times a deleterious mutation proved to be a prerequisite for EQU – as it often seems to be the case for complex features of biological organisms – and that – indeed, as it happens in biological evolution – complexity of the functions that are present in the ecosystem increases over time (cit., 141-143).

The issue of validating this experimental method appears however now quite tricky. How to judge if AVIDA is "successful" in modelling the real world behaviour of appearance of complex features? The problem is not only that the scientists have insufficient knowledge of the phenomenon that is being modelled: the problem is rather that it is very difficult to evaluate to which extent the behaviour of the model depends on the built-in assumptions. For example, most of the interesting results that AVIDA shows are ultimately depending on the fact that, as we noted earlier, the programs' ability to perform operations is rewarded at every level of complexity, at a measure exponentially increasing as complexity increases. In other words, the digital organisms in AVIDA develop within an environment that actively promotes emergence of complex features: the process which is at the origin of complexity in AVIDA is an artificial one and it is very likely not the same at work in the natural systems[1]. Indeed, Lenski and his colleagues tested a different model in which only the most complex function was rewarded, while none of the intermediate steps provided advantage: in this setting no population evolved EQU, showing that the reward strategy is quite crucial to the model's results. From what we have been arguing, it seems controversial to think that evolutionary algorithms like AVIDA do recapitulate this particular causal processes at work in natural evolution. This statement however is by no way demeaning the value of this computer model for illuminating the particular question for which it was conceived: namely, to assess whether complex features can be built upon simpler bricks by an unconstrained search process and in absence of a pre-existing plan.

The analysis of AVIDA's results provided above allows us to eventually point out some conclusions. The value of computer experiments is that they represent an unambiguous test: the program behaves deterministically, and will evolve according to the rules that the programmer scientist has set. A consequence of this is that computer experiments are naturally optimal tools for hypothesis disconfirmation. In this case, the adequacy of a theoretical model of a phenomenon can be assessed by building a computer model based on it that incorporates the key causal factors: if the computer model fails to reproduce the relevant aspects of the modelled phenomenon, this is a sign that the starting hypothesis has to be revised. On the other hand, making positive assertions from a computer model is less straightforward and, as we have seen, a move not always granted: due to the entanglement of experiment and theory which characterizes simulations, the detection of artefacts and the validation of the results are generally quite problematic. Therefore, the use of simulations appears to be a good strategy for testing particular, point-like hypotheses,

---

[1] On this, see for example Bedau (2009). The idea that evolution promotes complexity was famously challenged by Gould (1996).

while their use for general level investigations of the behaviour of a complex system appears to be epistemologically shakier.

5 Conclusion

In this work I have analysed the use of computer simulation in evolutionary biology. In evolutionary studies simulations with digital organisms allow, like experiments with model organisms, to study a particular causal link of interest in isolation. However, simulations rely on their theoretical content much more than model organisms do. The peculiarity of numerical simulations is indeed their being hybrid between experimental systems which are *observed* and model systems which are *built*.

The advantage of computer experiments is their extreme flexibility: they can be manipulated at ease in a way which natural systems cannot. This in turn, together with their proximity to theoretical constructs, makes them excellent tools for hypothesis refutation. On the other hand, making positive assertions from computer experiments is more delicate, since the artificial nature of the system makes validation and artefact detection problematic. For this reason, extreme care should be put in every case in checking the built-in assumptions and the extent to which the conclusions are possibly built into the program.

References

Bedau, M.A. (1999): "Can Unrealistic Computer Models Illuminate Theoretical Biology?" in Wu, A. (Ed.), *Proceedings of the 1999 Genetic and Evolutionary Computation Conference Workshop Program*. San Francisco, Morgan Kaufmann, pp. 20-23.

Bedau, M.A. (2009): "The Evolution of Complexity", in Barberousse, A., Morange, M. and Pradeu, T. (eds.), *Mapping the Future of Biology: Evolving Concepts and Theories*. Berlin, Springer Verlag, pp. 111-132

Fox Keller, E. (2003): "Models, Simulations and 'Computer Experiments'" in Hans Radder (ed.), *The Philosophy of Scientific Experimentation*. Pittsburgh, University of Pittsburgh Press, pp. 198-215.

Franklin, A. (1986): *The Neglect of Experiment*. New York, Cambridge University Press.

Gould, S.J. (1996): *Full House: The Spread of Excellence from Plato to Darwin*. New York, Harmony Books.

Hartmann, S. (1996): "The World as a Process: Simulations in the Natural and Social Sciences", in Rainer Hegselmann (ed.), *Modelling and Simulation in the Social Sciences from the Philosophy of Science Point of View*. Dordrecht, Kluwer, pp. 77-100.

Lenhard, J. (2007): "Computer Simulation: The Cooperation Between Experimenting and Modeling". *Philosophy of Science*, 74, pp. 176-194.

Lenski, R.E.; Ofria, C.; Pennock, R.T.; and Adami, C. (2003): "The Evolutionary Origin of Complex Features". *Nature*, 423(6936), pp. 139-144.

Ofria C.; Adami C. (2002): "Evolution of Genetic Organization in Digital Organisms", in Landweber, L. and Winfree, E. (Eds.) *Evolution as Computation*. New York, Springer, pp. 167.

Peck, S. (2004): "Simulation as Experiment: a Philosophical Reassessment for Biological Modeling". *Trends in Ecology & Evolution*,19(10), pp. 530-534

Ray, T.S. (1991): "An Approach to the Synthesis of Life". *Artificial Life II*, 11, pp. 371-408.

Taylor, T.; and Hallam J. (1997): "Studying Evolution with Self-Replicating Computer Programs", in Husbands, P. and Harvey, I. (Eds.), *Proceedings of the Fourth European Conference on Artificial Life (ECAL97), Brighton*, Cambridge, MIT Press, pp. 550–559.

Winsberg, E. (1999): "Sanctioning Models: the Epistemology of Simulations". *Science in Context*, 12(2), pp. 275-292.

The Limits of Mechanistic Explanation in Molecular Biology

Fridolin Gross
European School of Molecular Medicine (SEMM), Milan,
IFOM – Istituto Firc di Oncologia Molecolare, Milan,
and University of Milan
fridolin.gross@ifom-ieo-campus.it

1 Introduction

Molecular biology has been undergoing considerable changes in the past years. The ability of sequencing whole genomes has triggered the development of sophisticated new research tools that broaden the ways in which biologists approach their subject matter and confront them with a previously unknown richness of data. In order to cope with the complexities unraveled by these techniques, concepts and methods stemming from other disciplines – such as mathematics, physics, computer science or statistics – are increasingly incorporated into biological research. Instead of playing the role of mere tools, these concepts are now likely to shape in an essential way how biologists conceive of living things. Consequently, this development may also change what is considered to be a good explanation in biology.

Up to now, philosophers of science have mainly tried to capture explanations in biology in terms of mechanisms (e.g. Bechtel and Richardson 1993; Glennan 1996; Machamer et al. 2000). The framework of mechanistic explanation is closely linked to the highly successful research program that has guided molecular biologists throughout the second half of the twentieth century. Undoubtedly, the philosophical analysis of the concept of mechanism has contributed a great deal to our understanding of the life sciences and their relationship to other sciences. However, it is not clear whether this framework remains adequate in the light of the aforementioned changes.

The aim of this article is to discuss, with the help of a concrete example, how the paradigm of mechanistic explanation is challenged by recent findings in molecular biology. The example I have chosen makes use of the theory of dynamical systems to elucidate the processes of cellular development. The main idea behind it is to conceive of cell fates as attractor states in a high dimensional state space. My basic claim is that this view does not nicely fit within the mechanistic paradigm. From this diagnosis, I will conclude that the mechanistic framework has to be complemented by a different perspective on explanation in the life sciences. Moreover, the given example points to a more general framework in which mechanistic explanation can be integrated.

In what follows I first briefly discuss the concept of mechanistic explanation and highlight some of the features which are important for my further analysis. Then I turn to the chosen example and give some biological and technical background. After that I will discuss how this example relates to the mechanistic view.

2 Mechanistic Explanation

There are several reasons for the interest philosophers of science have taken in the concept of mechanism. On one hand, it can be seen as a reaction to the inadequacy of nomological explanations in the biological disciplines: Attempts at reducing explanations in the life sciences to "first principles" of the more basic sciences like physics and chemistry seem to have failed. It has become increasingly clear that the particularities of living systems, that is, their contingent aspects with respect to the known laws of the exact sciences, have to be given an essential role in biological explanations. Aside from that, mechanistic explanation seems to be a concept that fits well with what biologists normally offer as explanations in their scientific publications. The activity of molecular biologists has been characterized as «explaining types of phenomena by discovering mechanisms» (Wimsatt 1970, 67). In this way, the philosophical debate on mechanisms mirrors the general development of biology in the 20th century from a purely descriptive to an explanatory discipline.

In one of his early articles, Glennan (1996) envisions the mechanical view as an alternative to regularity theories of causation. He proposes that we should speak of a causal link between two events if we are able to describe a mechanism that links them. The description of a mechanism is satisfactory and accounts for the phenomenon if it includes all the relevant parts and the direct interactions between them. Even though he is still convinced of the importance of fundamental laws of nature, Glennan argues that, at higher levels, regularities are to be explained by mechanisms and not, as in nomological accounts, mechanisms by regularities.

Machamer et al. (2000) give a similar characterization, but they highlight the productive aspect of mechanisms. They use a concept of *activity* instead of *interaction* to stress that a mechanistic account does not only describe the changes underlying a certain behavior, but also makes intelligible what *brings about* this change. Mechanistic descriptions have explanatory power because we can literally *see* how the phenomenon is produced. Machamer et al. try to capture the concept of mechanism in the following words:

> Mechanisms are entities and activities organized such that they are productive of regular changes from start or set-up to finish or termination conditions. (Machamer et al. 2000, p. 3.)

Therefore, apart from their active nature, two further aspects of mechanisms seem to be crucial: organization and regularity. It is not enough that we identify the relevant parts and their activities: an essential aspect of mechanistic explanations lies in the description of how these parts have to be organized to give rise to an ordered sequence of events producing the regular behavior we want to account for.

In a similar vein, Glennan (2005) discusses the difference between the mechanistic view and the more general framework of the semantic view of scientific theories (e.g. Suppe, 1989). For this purpose, he explicitly distinguishes between mechanisms as entities existing in the real world, and our representations of mechanisms which he refers to as *mechanical models*. According to Glennan, a mechanical model can be understood as a special instance of a state space model. However, in the general case of a state space model there is considerable freedom in choosing the state variables as long as they fully characterize the state of the modeled system. In the mechanical model, on the other hand, the state variables have to directly correspond to the internal working processes of the system. He gives the example of a watch whose state can be fully characterized by giving the positions of the hour hand and the minute hand. Two variables are therefore enough to characterize the behavior of the watch in a general state space model. In order to count as a mechanical model, however, the description will have to include all the cogs and springs etc. inside the watch that produce the observed behavior of the hands.

In general, proponents of the mechanistic view seem to agree that mechanistic explanations do not have to go to the deepest possible level of description. Mechanisms at a lower level may figure as parts of higher level mechanisms and can then be explained separately. Moreover, most disciplines *bottom out* at a certain level (cf. Machamer et al., 2000), that is, they consider entities and their interactions below that level as unproblematic or irrelevant. For example, molecular biologists are usually not interested in subatomic aspects of matter such as the detailed composition of nuclei.

Advocating mechanistic explanations in biology implies a certain perspective on living organisms in general. In order to lend itself to a mechanistic explanation, the system of interest as a whole has to be organized in a certain way. Not any complex system allows us to single out well defined behaviors and explain them by referring to an isolated subset of parts. Moreover, even if a system exhibits such a modular structure by nature, it is not obvious that the modules in which we break it down will be simple enough to allow for mechanistic descriptions. If a living system is to be fully explained by means of mechanistic explanation, it must ultimately be organized in a hierarchy of nested mechanisms. The mechanistic framework therefore presupposes a fundamentally modular structure of biological systems, and it is grounded on the hope that we can make the mechanisms underlying this structure intelligible.

3 A Globalist Dynamical Perspective on Gene Regulatory Networks

3.1 Biological Background

In multicellular organisms all types of cells arise in a number of differentiation steps from the same kind of undifferentiated cell and, therefore, all of them share the same genome. However, different cell types show very different properties and are able to perform completely different tasks within the organism. These differences arise from different ways in which the genome is regulated in each cell type. In the course of its development each cell undergoes a number of discrete cell fate decisions in which it acquires certain phenotypic characteristics and loses others. What is remarkable about these differentiation steps, is the reliability with which they occur and the stability with which the cellular phenotype is maintained. Until recently, this was explained as the irreversible switching on or switching off of cell specific pathways by means of so called *epigenetic marks* such as DNA methylation and modifications of the chromatin structure, the idea being that certain regions of the genome are simply not accessible to the transcriptional machinery. These changes were thought to be induced by certain very specific external stimuli. However, recent findings suggest that the epigenetic marks are not as stable and specific as previously thought. Many of the changes to the genome are reversible and highly dynamic. The plasticity of cellular phenotypes and the successful reprogramming of somatic cells further put into question the traditional view of irreversible cell fates (cf. Huang, 2009).

An alternative perspective on differentiation and cell fates has recently entered the domain of experimental molecular biology. It is the idea of understanding cellular phenotypes as attractors in a high-dimensional state space. This view can explain the stability of cellular phenotypes, the reliability of developmental paths, as well as phenomena of cellular plasticity in a rather natural way. The general idea, however, is not new and was already formulated, at least in rudimentary forms, by people like Waddington (e.g. Waddington 1956), and has been treated formally by Kauffman (e.g. Kauffman 1969). New methods, in particular microarray technology, now provide a way to connect the attractor view to experimental practice.

3.2 Technical Background

The basic assumption that makes the state space view accessible to experiment is the idea that the state of a cell can be approximated by the expression levels of the genes participating in the network of gene regulation. This network is composed by all the genes whose products influence the expression of other genes either indirectly by means of protein-protein interactions or by directly acting on the genome as transcription factors. A particular state of this network can formally be expressed as a vector $S=(x_1,x_2,\ldots,x_n)$ where each of the x_i corresponds to the expression levels of one gene. The set of all possible combinations of expression values will then span a high-dimensional state space in which each point represents a (theoretically) possible state of the cell. The dimension of this space can be approximated by the number of transcription factors; for the human genome this number is estimated to be of the order of 1000 (e.g. Farnham, 2009).

Most states in this space, however, will be unstable because the expression value of each gene is subject to constraints imposed by the expression levels of other genes. To give a simplified example: it may not be possible for gene A to be co-expressed with gene B because the protein that B codes for inhibits the transcription of A. Therefore, a state in which both A and B are expressed at high levels is unstable and will quickly change towards a state in which expression of A is low. In general, it is expected that there is only a comparably small set of stable states in the state space. A cell that is not in one of the stable configurations will converge to one of stable states, which is why they are called *attractors*. But changing the transcriptional state of the genome along its path through the state space, the cell will also change its phenotypic properties. For this reason, the state space can also be interpreted as a *phenotypic space* which implies that attractor states correspond to stable cell fates. Theoretical analysis has shown that large networks, of the kind we are interested in here, will allow for only a relatively small number of attractor states (Kauffman, 1969; Huang, 2004).

The level of abstraction of the state space representation allows researchers to go beyond the mere architecture of the network, and to investigate its dynamics by studying the temporal behavior, $S(t)$, of the transcriptional state. The cell, represented at a certain moment in time by a single point in the state space, will move along a trajectory which is dictated by the dynamical relationships holding between all the genes composing the network. The attractor view, therefore, can be regarded as a mathematical foundation of Waddington's famous metaphor of the *epigenetic landscape*, in which the process of cell differentiation is compared to the movement of a marble down the slope of a rugged landscape (e.g. Waddington, 1956).

3.3 Experimental Methods: DNA-Microarrays

Scientific and technological advances in the course of the past two decades have enabled the development of experimental methods that allow biologists to monitor the behavior of molecules inside the cell on a large scale. One of these, microarray technology, dates back to the mid-1990s and has since then revolutionized the analysis of gene expression. A DNA Microarray is a glass chip that contains thousands of different DNA fragments on its surface, each corresponding to a particular gene. Each type of fragment can be localized by its exact position on the chip. In a microarray experiment messenger RNA (mRNA) extracted from a biological sample is copied into DNA (cDNA) and then allowed to hybridize with the fragments on the chip. The cDNA is labelled with a fluorescent probe to enable the detection of the positions where the fragments have bound to DNA from the sample by means of a laser microscope. The intensity of the emitted fluorescent light is proportional to the number of hybridizations, which in turn provides a measure of expression of the corresponding gene (see e.g. Alberts et al., 2008, 574-575). In this way the chip allows researchers to monitor the expression levels of thousands of genes simultaneously. Moreover, time-series data of the transcriptional state can be obtained by repeated measurements. In this way the previously abstract concept of the state vector $S(t)$ turns into a directly observable quantity. Obviously, the information obtained in a microarray experiment is initially just a large set of numbers which in order to provide biological understanding has to be further analyzed. There are statistical tools, some of which developed in completely different contexts, that facilitate this work. Methods such as principal component analysis and self-organizing maps help to represent the data in a way that makes them intelligible.

In the following, an example of an experiment that has been carried out by means of these methods will be described in more detail to illustrate the utility of the interpretation of cell fates as attractors.

3.4 Evidence for the Attractor View: Convergence of Trajectories

As mentioned already, the notion of an attractor state is one of the main concepts within the dynamical perspective. Attractor states have two important properties: First, they are fixed points, that is, once the system is in an attractor state it will remain there forever (provided that the system is not subject to external perturbations). Second, attractor states are stable, which means that trajectories starting from states in the vicinity of such a state will eventually converge towards it.[1] The set of all states that converge to a particular attractor state is called its *basin of attraction*.

In the traditional view, cell differentiation is a deterministic process in which the final state of a cell is the result of a well defined chain of molecular events. If we conceive of cell fates as attractor states, however, the picture changes considerably. The final state of differentiation is not primarily seen as the end of a particular chain of molecular events, but as a state that is simply more stable than the states in its vicinity. Just like in the case of a marble rolling down a rugged slope, many different trajectories to reach the stable state are possible, and, therefore, there are many different chains of molecular events that lead to the same final state. In theory this sounds compelling, but is it possible to detect this kind of behavior in living cells?

---

[1] Note that the two properties are independent: there can also be unstable fixed points.

21

I will now turn to the description of a result reported in Huang et al. (2005) that gives first evidence of the adequacy of this view. The experiment investigates the differentiation of neutrophils. Neutrophils are white blood cells that are derived from a particular type of progenitor cells, so called *promyelocytic* cells. Notably, these progenitor cells can be induced *in vitro* to differentiate into neutrophils by a variety of different stimuli. From the perspective of the attractor view, the state of the progenitor cell is destabilized by the external perturbation and enters the basin of attraction of the neutrophil state. The crucial idea behind the experiment is that different stimuli will induce the cell to differentiate towards the same neutrophil attractor state via different trajectories. In the case described, neutrophil differentiation is induced by two biochemically distinct stimuli, and the transcriptional state of the differentiating cells is monitored over time using microarrays. Visualizations of the transcriptional state of the cell in the form, for example, of self-organizing maps clearly show how trajectories initially diverge to different regions of the state space, but eventually converge to virtually identical expression patterns (see Huang et al. 2005, fig. 1).

It seems awkward to claim that experiments such as the one described *prove* the existence of high-dimensional attractor states in the gene regulatory network. Attractors, to begin with, are abstract entities that only exist within a given theoretical model. Whether or not they turn out to be an adequate concept for our description of biological systems depends on their potential to improve our biological understanding. What should be considered, however, is that the attractor view represents a perspective that is taken seriously by some biologists themselves and should probably be taken seriously by philosophers of biology as well. In the next section I will, therefore, investigate whether the philosopher's framework of mechanistic explanation is able to accommodate the prospect offered by the attractor view.

4 State Space representation and the Mechanistic View

The state space representation of the gene regulatory network describes a complex system with parts and interactions that are in principle well defined and whose organization gives rise to well defined behaviors. At first glance it seems, therefore, that it meets the criteria for mechanistic explanation quite well: A biological phenomenon (cell differentiation) is explained by referring to an organized systems of parts (genes, mRNA, proteins etc.) and their activities or interactions (transcription, suppression, activation etc.). However, difficulties arise when it comes to explicitly showing how the phenomenon is brought about in terms of the mentioned parts and interactions. The "all-encompassing eye" of high-throughput experimentation reveals cell differentiation as a process in which large parts of the whole genome are simultaneously involved. To monitor this process in the state space of gene expression means to completely abstract from the detailed causal structure of gene regulation. Individual genes are treated anonymously as components of a state vector, that is, as numbers. Their names and specific functions are not of interest, and no attempt is made to explicitly model the astronomic number of interactions between them. As a result the overall meaning of the outcome of a high-throughput experiment changes as well. It is more than just carrying out a large number of individual measurements on different molecules in parallel; rather, it is understood as one measurement of the state of the system as a whole. Consequently, there is no structural or functional decomposition in the mechanistic sense, and the phenomenon of cell differentiation is not explained by referring to components with specific properties that contribute to an overall activity of "differentiating". Instead, the process is accounted for by observing that the differentiated state corresponds to the most stable configuration.

The existence of different but converging trajectories in the state space seriously threatens the possibility of giving a satisfactory explanatory mechanistic account of cell differentiation phenomena. As we have seen, during these processes the same terminal state may be reached via an indefinite number of different trajectories, each corresponding to a qualitatively different chain of molecular events. Strictly speaking, each of these has to be given a different mechanistic description. Thus, even if it turns out that we find manageable mechanistic descriptions at the gene level for every single cascade of events, they will not account for the observed higher level regularity in cellular behavior. We are, therefore, confronted with an instance of regularity resisting mechanical explanation in precisely the sense that Glennan had already anticipated as a theoretical possibility:

The weak reading [of the mechanical explicability of non-fundamental laws] allows for the possibility that there are higher level laws, every instance of which must be explained by a different mechanism, perhaps even by a mechanism of a radically different kind. In such cases, the laws in question would not genuinely be *explained* by reference to these mechanisms, because nothing can be said about how the *type* of lawful behavior is produced by mechanisms. Such strongly irreducible laws would, like fundamental laws, resist mechanical explanation, but would, unlike fundamental laws, supervene on lower level mechanisms. (Glennan 1996, p. 62)

It has to be stressed that the possibility of this scenario does not entail that cell differentiation ontologically distinguishes itself radically from other biological processes. Rather, it suggests that the framework we are using to explain such processes turns out to be unsatisfactory when complexity reaches a certain level.

Does this mean that some biological processes are just so complicated that we have no chance in ever reaching understanding at the molecular level? Not necessarily. The attractor view, apart from disclosing the limits of mechanistic explanation, might also point toward an alternative, or better, a more general framework of explanation in molecular biology. As we have seen, the concepts of the abstract theory of dynamical systems can help us to grasp better certain features of biological processes also in the absence of knowledge about the detailed molecular events. Even though the investigation of developmental trajectories in the state space does not fulfill Glennan's requirements of a mechanical model, and is rather the equivalent of representing the clock's state by the position of is hands, it seems to carry big potential to improve our biological understanding in the future. It might, moreover, be useful to adopt a more general, global perspective on biological systems even in cases where strong mechanistic regularities do exist. After all, descriptions of mechanisms only explain the phenomena they are supposed to explain and do not necessarily explain their own reliable and regular working in the bigger context of the system they are embedded in.

5 Conclusion

The goal of the present article was to present a concrete example of biological research for which the philosophical framework of mechanistic explanation seems insufficient. Microarray experiments have shown that terminal states in cell differentiation show properties of high dimensional attractors. This suggests that for biological systems, such as the cell, the global characterization of dynamical properties might often be more useful than the description of causal events at the molecular level. In fact, as we have seen, it may often turn out that one behavior corresponds to a unmanageable number of different underlying mechanisms.

It is not impossible that future experiments will reveal that, after all, the set of differentiation trajectories is restricted to a small number of types, and that each type can be characterized by some key molecular events. In the meantime, however, we cannot exclude that the global perspective described here might be the only way to capture the regularity and robustness involved in cellular development. The attractor view, even in its purely descriptive form, already improves our understanding of cellular development, leads to interesting hypotheses and suggests further experiments. Instead of insisting on one single framework of explanation, philosophers should investigate in detail how different explanatory strategies interact in practice, especially in interdisciplinary endeavors such as contemporary research in molecular systems biology.

References

Alberts, B.; Johnson, A.; Lewis, J.; Raff, M.; Roberts, K.; and Walter, P. (2008): *Molecular Biology of the Cell*. New York, Garland Science, 5th edition.

Bechtel, W.; and Richardson, R.C. (1993): *Discovering Complexity: Decomposition and Localization as Scientific Research Strategies*. Princeton, Princeton University Press.

Farnham, P.J. (2009): "Insights From Genomic Profiling of Transcription Factors". *Nature Reviews Genetics*, 10, pp. 605–616.

Glennan, S. (1996): "Mechanisms and the Nature of Causation". *Erkenntnis*, 44, pp. 49-71.

Glennan, S. (2005): "Modeling Mechanisms". *Studies in History and Philosophy of Science Part C: Studies in History and Philosophy of Biological and Biomedical Sciences*, 36, pp. 443-464.

Huang, S. (2004): "Back to the Biology in Systems Biology: What Can We Learn From Biomolecular Networks?". *Briefings in Functional Genomics and Proteomics*, 2, pp. 279-297.

Huang, S. (2009): "Reprogramming Cell Fates: Reconciling Rarity with Robustness". *BioEssays*, 31, 546-560.

Huang, S.; Eichler, G.; Bar-Yam, Y.; and Ingber, D.E. (2005): "Cell Fates as High-Dimensional Attractor States of a Complex Gene Regulatory Network". *Physical Review Letters*, 94, pp. 128701-1 – 128701-4.

Kauffman, S. (1969): "Homeostasis and Differentiation in Random Genetic Control Networks". *Nature*, 224, 177-178.

Machamer, P.; Darden, L.; and Craver, C. (2000): "Thinking About Mechanisms". *Philosophy of Science*, 67, pp. 1-25.

Suppe, F. (1989): *The Semantic Conception of Theories and Scientific Realism*. Chicago, University of Illinois Press.

Waddington, C.H. (1956): *Principles of Embryology*. London, Allen and Unwin Ltd.

Wimsatt, W. (1972): "Complexity and Organization", in K. F. Schaffner, & R. S. Cohen (eds.), *PSA 1972: Proceedings of the Philosophy of Science Association*. Dordrecht: Reidel, pp. 67-86.

Health and Disease between Biology and Bioethics

Fabio Lelli
University of Bologna
fabio.lelli@unibo.it

1 Bioethics and philosophy of science

The relationship between bioethics and philosophy of science is not simple. Most of the scholars involved in bioethics are coming from moral philosophy (more rarely from the philosophy of politics or philosophy of law) or from medicine. Philosophers of science and hence philosophy of science in bioethics are lacking. Let me quote Giovanni Boniolo:

> I think it's necessary that a bioethics' scholar should be, first of all, a philosopher of biology, that is, a philosopher of science interested in biological sciences, and subsequently an intellectual with a well-grounded knowledge of moral philosophy. (Boniolo 2003, p. 365).

Failing to keep all the relevant disciplines together and underestimate the epistemological issues involved could lead to contradictions and ambiguities. For instance, as I wish to show in this paper, if we really think of "health" (and its correlative "disease") as a purely scientific or purely moral concept, the net result would be a misrepresentation of all the real questions that bioethics should address.

2 What is the role of "health" in bioethics: Norman Daniels and the liberal philosophy

Bioethics relies upon an inner tension: one of its ubiquitous claim is to build a solid basis in order to resolve ethics dilemmas generated by biological science, but at the same time every bioethical stance is grounded in a specific moral and political view that is not universally shared; therefore is almost impossible to reach an universal agreement.

To address this dichotomy, *liberal* bioethics (Rawls, Dworkin, Daniels, and Charlesworth) does not include a normative model of "good life", but tries to establish the *necessary conditions* in order to allow everyone to pursue his own "project of life". What the State must ensure, with the institutionalised use of force, is the *equality of opportunity*. It is therefore necessary to take into adequate account the material conditions of life of each individual, and among these it is reasonable to include *health*. For this reason, Norman Daniels insists on the defence of health care: if we cant match with our "normal functioning" (from a biomedical point of view) we will not have access to the "the range of opportunities we would have, were we not ill or disabled, given our talents and skills" (Daniels 1998, p. 316; see also Daniels 1985, and Daniels 2001).

Daniels takes for granted that this "normal functioning" —which, according to him, is equivalent to health— is a scientific concept, and as such is beyond any possible recrimination and accusation of bias. On the epistemological side, this account refers to the definition of health which has been formulated by Christopher Boorse.

3 The "biostatistical theory" of health

Boorse's position, named "biostatistical theory", was developed through various interventions since 1975 (Boorse 1975; Boorse 1977; Boorse 1997). In this approach, health coincides with the *absence of diseases*, whereas a disease involves a reduction of one or more functional capacities below their typical efficiency caused by an internal or external environmental agents. "Typical" means the conformity to the *model of the species* as it is shaped during its evolutionary history, and this model is described by identifying the functions that statistically contribute to the purposes indicated by the evolution theory: survival and reproduction. This perspective is not epistemologically naive: Boorse points out that statistical data in themselves *do not say anything unless* they have been read through the grid of evolution theory. For example, being red-haired is certainly more statistically rare, but it has no influence on the model of the human species with respect to its evolutionary purpose. The same can be said for some classical parameters that have historically been read as the equivalent of health, such as *homeostasis*: actually some events that disrupt the homeostasis, such as childbirth, are essential for the purposes of evolution of the species. The aim of this theory is to be a value-free teleological perspective, since the purposes, the choice of which justifies the whole building, are taken from the theory of evolution, limited to the survival and reproduction, and should be objective and valid for all.

4 Boorse criticism. Nordenfelt and Engelhardt

Boorse's theory is based on the purposes of a living being according to the evolution theory, but these purposes does not always correspond with the goals pursued by human beings as "persons", namely in terms of ethics and political philosophy. The main goal pursued by the *citizens*, according to a liberal philosophy, as mentioned, is to develop their own life plan, and it is far from clear that this coincides with the two biological purposes, or even that the biological purposes are relevant to its implementation.

As opposed to Boorse's account, there are radically different theories of health which are not value-free, since their authors would argue that the concept of health is essentially moral. In particular, the theory advanced by Lennart Nordenfelt is based on a complex analysis of the concepts of "action" and "happiness". Here a quote of his final definition of health should be sufficient for our purpose: "A is completely healthy if and only if A is in a bodily and mental state such that, A has the second-order ability, given an accepted set of circumstances C, to realise all her vital goals" (Nordenfelt 2000, p. 162). The expression "second-order ability" means the ability to acquire a capacity after a certain training (e.g., not being able to play piano is different from the inability of being able to play the piano for some reason even after the necessary training), and the "accepted set of circumstances" are realistically normal conditions relating to some physical and social environment.

5 Some remarks to Nordenfelt's theory

The choice of a perspective focused on the goals of human beings in society is certainly more consonant with the view of bioethics we have considered in the first section, but this will rely on several parameters that can be interpreted in the most different ways: what does it really mean to have an ability of the second order? What are the "accepted set of circumstances" within a particular society? What about the "vital goals"? According to Nordenfelt's analytical proposal "the vital goals of A are the set of states which are such that their realisation is necessary and jointly sufficient for A's minimal long-term happiness" (Nordenfelt 2000, p. 163), where happiness is a point of equilibrium between the subject wants and the world as he finds it to be. One important achievement of this view is to make happiness and health (almost) independent from each other: from the definition proposed it should follow that one can be in perfect health even if he is unhappy, and vice versa being healthy doesn't grant an enjoyable happiness; these outcomes are in step with a very basic insight of the liberal philosophy.

What about those "standard conditions"? For Nordenfelt our wishes could be "reasonable" given the external circumstances; e.g., our strong will to build a house at the top of Himalaya is not reasonable. But examples of this kind are not problematic: what if I do need to assume a very expensive medicine for the rest of my life in order to be able to walk with my own legs? Resources are always scarce, and someone has to decide whether the context makes my request reasonable or unreasonable, and even the definition of happiness as an "equilibrium" implies a judgement upon the state of the external world.

Politicians are in charge for these decisions, thus health care depends on their planning, leaving behind the medical science. Nordenfelt admit this conclusion in *On the Nature of Health* (Nordenfelt 1987, chaps. 5 and 6). The real threat is that in this framework biomedical conditions could be, and this would be highly unreasonable, ruled out.

The conclusion is that the formulation of vital goals is in charge of policy makers and of the authorities responsible for the formulation of health programs; the clinician and the scientist must work in light of these concepts. [5]

In short, whereas in Daniels, via Boorse, science (evolution theory) *injects its own ideals* into political decision-making —and in so doing also covers much evaluative ground, but without this being stated outright— for Nordenfelt science ought to be purged from the "control room" and *made to depend on* political decision-making, in a process where strict medical-biological considerations need not have any role.

6 The advantages of a biomedical concept of health

A much-influential view in bioethics has been that of H.T. Engelhardt (for instance, see Engelhardt 1986), for whom the idea of health can sensibly be elucidated only in the context of a given "moral community", for health is a matter of values, in that values inform and indeed shape our understanding of health. Thus *health* can only be defined within a social group that shares basic moral values (a group such as the moral community of Catholics, atheists, Marxists, and so on): it follows that no such definition is available in any outside setting, especially in the state. But what also follows is a bioethics which places personal autonomy ahead of any other consideration, and for which, accordingly, no government action could legitimately undermine the minimum conditions for the existence and coexistence of moral communities.

If health is argued to depend in an essential way on personal and social evaluation, the idea that the state should actively promote health consequently comes out weakened, because if health is not a scientific concept, then it cannot be included in the basic structure that everyone will accept.

This is the risk of theories like Nordenfelt's and Engelhardt's: their theories may have a deeper moral grounding, but in this way they systematically underrate the biological factors involved.

7 Haemophilia, for example

It may be that the two theories just briefly outlined share the same inherent limit: they both conceive medicine and human values as closely independent (Lelli 2007). This can be appreciated by taking as an example a serious chronic disease, such as haemophilia, and considering it from a more nuanced epistemological standpoint, one that no longer relies on a strict distinction between facts (biology) and values (morals), on the premise that the human being is a biological organism embedded in a world of values, such that *both* factors (biology and values) have to be taken into account for a proper understanding of the human being.

Haemophilia is a bleeding disorder owed to chronic alteration of the gene that governs the synthesis of a protein essential for clotting: factor VIII in haemophilia A, factor IX in haemophilia B. It has a monofactorial genetic cause and it affects about one in ten thousand newborns. For the most part its effects are not totally debilitating. Indeed, it might be argued that the disease simply *diminishes* the organism's normal capacity. A very effective prophylactic technique is available today that involves the infusion of an artificially synthesised concentrate of factor VIII —a technique that allows the patient to "normally" engage a wide range of activities, including sports.

That is to say that with the prophylactic technologies currently available, a haemophiliac can do anything a non-haemophiliac can do. Unfortunately, prophylaxis is extremely expensive, and without an entitlement program, only few patients could afford the treatment. But it is also true that most haemophiliacs can lead a normal life even without the infusion, except that they would have to avoid certain activities, like sports. So, what would Nordenfelt's "accepted conditions" be for them? Would that mean not being able to enjoy activities which most other people can engage in (and which may be essential to their life plan)? If so, on Nordenfelt's definition, is haemophilia a disease only in those societies where prophylaxis is not covered by health insurance or by a national health service?

On the other hand, the biostatistical definition of *health* does not capture the particularities of this disease, especially its ability to reduce the range of available options: the definition fails to capture anything that should fall outside the purposes of biological evolution.

So what kind of a disease is haemophilia? At the microbiological level, it all comes down to interactions between molecules: an abnormality in the DNA produces in small quantities another molecule (RFV VIII) having a diminished capacity for molecular reactions linked to blood clotting. Such explanations have been described as "biochemical lesion" and "molecular pathology", two terms coined in a seminal article by Pauling and Zucherkandl first published in 1949 (Pauling & Zucherkandl 1962). The lesson to be learned, however, is not that this disease is a condition affecting certain molecules, because it is not the molecules themselves that can be described as "sick": the disease rather lies in the interaction *between* molecules. The authors also note that what we now call "molecular diseases" are phenomena that in earlier times have been a basis of evolution, in the effort to find ways to adapt to the environment in which the body is living. In other words, even the functional features we now have were once labelled "molecular diseases". But the point is that there are different types of intermolecular relationships suited to a variety of environments. This is a very complex chain of relationships, and it does not stop at a microbiological level, either: through the same kind of reasoning, we can extend the view from molecules to cells, from cells to organs and individuals, and from individuals to their environment. And this environment includes the whole of human society. In this sense, a disease is to be viewed as an inconsistency between and individual and his or her environment: its causes must always be traced to multiple interacting levels, and we cannot in our analysis draw a line of separation between the biological and the social.

One explanation of these complex interactions can be found by looking to evolutionary medicine (for a broad introduction, see Zampieri 2009). which has the advantage of *coupling*, rather than counterposing, the biological element with the social (and hence with morals and politics).

Health, on this view, can be described as an adaptability to the environment sufficient to enable us to each lead a life according to our own aims and values. And our environment will always include, on one hand, the traits of our metabolism and the specific nature of any disease, which each body lives in a unique way, and on the other hand, the social setting and economic situation we are in, interacting with our molecules and with our values and goals, both essential sources of meaning.

References

Boniolo, G. (2003): "Filosofia della biologia: che cos'è?", in L. Floridi (ed.), *Linee di ricerca*, SWIF, pp. 350-393.

Boorse, C. (1975): "On the Distinction between Disease and Illness". *Philosophy and Public Affairs*, 5, pp. 49-68.

Boorse, C. (1977): "Health as a Theoretical Concept". *Philosophy of Science*, 44, pp. 542-573.

Boorse, C. (1997): "A Rebuttal on Health", in J.M. Humber and R.F. Almeder (eds.), *What is Disease?*, Totowa, Humana Press, pp. 3-134.

Daniels, N. (1985): *Just Health Care*. Cambridge, Cambridge University Press.

Daniels, N. (1998): "Is There a Right to Health Care and, if so, what does it encompass?", in H. Kuhse and P. Singer (eds.), *A Companion to Bioethics*, Oxford, Blackwell, pp. 316-325.

Daniels, N. (2001): "Justice, Health, and Healthcare". *The American Journal of Bioethics*, 1/2, pp. 2-16.

Engelhardt, H. Tristram jr. (1986): *The Foundations of Bioethics*. New York, Oxford University Press.

Lelli, F. (2007): "La medicina fra due mondi", *Discipline Filosofiche*, 12, pp. 323-340.

Nordenfelt, L. (1987): *On the Nature of Health. An Action-Theoretic Approach*. Dordrecht, Kluwer.

Nordenfelt, L. (2000): *Action, Ability and Health. Essays in the Philosophy of Action and Welfare*. Dordrecht, Kluwer.

Pauling, L.; and Zucherkandl, E. (1962): "Molecular Disease, Evolution and Genic Heterogenity," in M. Kasha and B. Pullman (eds.), *Horizons in Biochemistry*, New York, Academic Press.

Zampieri, F. (2009): "Origins and History of Darwininan Medicine". *Humana.Mente*, 9, pp. 13-38.

The Temporal Boundaries of Biological Species

Elena Casetta
University of Torino
elena.casetta@unito.it

1 Introduction

Ontological conventionalism is the view according to which some – or all – entities depend as their individuation on our conventions, in other words, they are the result of negotiation acts or, more widely, of social practices. In this paper I look at some consequences of this view for the ontological status of species in contemporary biology. Specifically, I argue that the temporal boundaries of species – hence, their persistence conditions – are partly conventional because they are individuated by social practices (where by 'social practices' I mean here scientific activities and, in particular, taxonomic practices).

The issue of the temporal boundaries of species is indirectly involved in the so-called Species Problem, which can be broadly summarized by means of three oppositions:

(i) *Nominalism vs. Realism*. For a realist, species are the 'natural joints' that the skilled Platonic butcher must discover and along which he must carve reality, whereas for a non-realist (for instance, a nominalist), species are just cognitive constructs, linguistic devices by means of which biologists – and the layman as well – dissect the external world.

(ii) *Pluralism vs. Monism*. For a monist, there is one and only one 'correct' species concept: the natural world has a structure of its own, and our biological taxonomy must mirror it. Several authors (e.g., Kitcher 1984, Dupré 1993, Ereshefsky 2001) are instead promoting a pluralistic approach, according to which there is no single correct species concept. A related issue is what may be called the *de facto pluralism*: it is a matter of fact that, in scientific practice, there is no unique and uncontroversial definition of species. Rather there are several species definitions, often mutually incompatible, simultaneously in use among biologists (26 species concepts in modern literature, see Wilkins 2009).

(iii) *Sets vs. Individuals*. On a traditional view, species are construed as sets whose members are individual organisms sharing an essential property or a cluster of properties. Some biologists (notably Ghiselin 1974 and Hull 1976) have argued instead that species are (large) individuals, for species evolve whereas sets (which are abstract entities) are necessarily static. On this view, an organism is therefore not a member of its species, but literally part of it.

In section 1, I state some assumptions that are needed to set out the general framework. In section 2, I take a closer look at the problem of defining the temporal boundaries of species, considering speciation and extinction, and I will show that several conventions are required in order to establish when a species originates and when it becomes extinct. Finally (section 3), I draw my conclusions: given that the temporal boundaries of species are, at least in part, determined not by biological facts but rather *drawn by convention following selected biological facts*, species identity is, at least in part, a matter of convention.

2 Assumptions

Three assumptions need to be made explicit before proceeding.

(a) The first assumption concerns the stance that, in what follows, I will adopt on the Species Problem. In particular, about (i) ontological conventionalism is generally considered to be more on the side of nominalism than realism (and realism seems in turn to be a better option than nominalism from an "operational" point of view—just consider: how could we explain the meaning of claims such as "Species are the fundamental units of evolution" or "*H. sapiens* is a different species than *D. melanogaster*" in a nominalistic framework?). Nonetheless, I think that conventionalism is compatible with both nominalism and realism.[1] About (ii) I won't take a stand in the dispute between monists and pluralists; I'll just accept the multiplicity of species concepts as a matter of fact (*de facto*

---

[1] I defend a form of conventional realism about species in Casetta 2009.

pluralism). Finally, about (iii) I will simply assume that species are *groups* of organisms, leaving it open whether groups should be construed as sets or as individuals.

(b) The second assumption is the acceptance of the principle that the ontological status of an object is determined by its boundaries. In particular, the persistence conditions of an object are determined by its temporal boundaries. Thus, if the boundaries of an object are conventional, the object itself is a conventional entity.

> If a certain entity enjoys natural boundaries, it is reasonable to suppose that its identity and survival conditions do not depend on us; it is a *bona fide* entity of its own. By contrast, if (some of) its boundaries are artificial – if they reflect the articulation of reality that is effected through human cognition and social practices – then the entity itself is to some degree a *fiat* entity, a product of our worldmaking. (Varzi 2011, p. 9).

(c) The third assumption concerns what is meant by 'convention' in the ontological conventionalism at issue in this paper. By 'convention' I mean "A norm which there is some presumption that one ought to conform to" (Lewis 1969, p. 99) that typically has been introduced to solve a coordination problem, namely a situation in which there are several ways agents may coordinate their actions for mutual benefit.

3 Temporal Boundaries

The temporal boundaries of a species are determined by two processes: speciation, namely the process by which a new species arises, and extinction, namely the process by which a species comes to an end. Evidently, both processes are dependent – as far as their characterization is concerned – on how we understand the concept of a species, and this requires a choice among several alternatives. Put differently: what I have called *de facto* pluralism requires that in order to make sense of the birth or the extinction of a species, X, it is necessary to make a choice and specify what type of species X is, i.e., according to which taxonomic school X is considered.

Take the most popular species concept, E. Mayr's *biological species concept*, according to which species are "groups of interbreeding natural populations that are reproductively isolated from other such groups" (Mayr 1970, p. 12). If we adopt this concept (that is, framing the point in Lewisian words: if we come to the conclusion that the biological species concept is the best solution to our coordination problem of finding a species definition) we will have certain criteria of speciation as well as certain criteria of extinction. We will say, for instance, that a new species arises from an ancestral species A when a population of A becomes reproductively isolated from the other populations of A. Analogously, we will say that a species A becomes extinct when its members are no longer capable of interbreeding, or something along these lines.

Mayr's concept – as well as the related definitions and criteria for speciation and extinction – clearly does not apply to organisms which do not reproduce themselves sexually, such as parthenogenetic animals and plants, many fungi, some green algae, many bacteria and parasites. (Contrary to Mayr, who considers them as uncomfortable anomalies, such organisms are a substantive part of the earth's biological diversity – see Van Dijk, Martens, Schön 2009.) Asexual organisms do not interbreed; nonetheless, they evolve, they are classified in species (although this has been questioned – see Maynard Smith and Szathmary 1995), they can go extinct (and their rates of extinction are generally considered higher than the rates of extinction of sexual species). As far as asexual organisms are concerned, the biological concept of species is not a good solution of the coordination problem raised by *de facto* pluralism. A better equilibrium is reached instead by means of the phenetic concept, or the ecological concept, which sort organisms into species on the basis of their overall similarity or niche sharing, respectively. Clearly, if a different concept of species is adopted, different criteria for establishing speciation and extinction will apply.

Summarizing: the criteria used by taxonomists for determining speciation and extinction are subsequent to the choice of a certain species concept, and this is the first convention needed in order to establish when a new species arises or when a species becomes extinct. The source of the coordination problem is what I am calling *de facto* pluralism (namely, the fact that biologists use several different species concept partitioning organisms in different and often mutually incompatible ways); the solution is an agreement – at least provisional – among biologists on what would be the best species concept in consideration of the type of organisms to be classified (e.g. the biological concept – or something similar – for sexual organisms; the phenetic or the ecological concept – or another concept that is not based on interbreeding – for asexual organisms.)

In what follows I will focus on sexual organisms and I will assume that a first convention has been established: if we are dealing with sexual organisms, the best way of classifying them into species is to look whether they interbreed or not.

3.1. Speciation

I consider only Mayr's standard model of speciation of sexual organisms, namely allopatric speciation, which is – as far as I know – uncontroversial. In allopatric speciation, geographically separated populations – groups of organisms considered members of the same species – evolve independently, resulting in speciation. Often this type of speciation occurs when some event isolates one or more populations from the original species' main body. According to this sketch, speciation seems quite clear and easy to recognize and detect. But – as Darwin already knew – speciation is a gradual process: a species is not formed by a punctual act of creation. Rather, when a sufficient number of appropriate changes has occurred (in a slow and gradual manner), one species is said to evolve into another species, or a portion of a species to originate a new species.

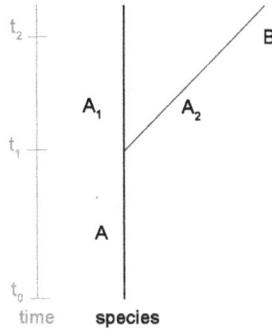

Figure 1

Consider a species A consisting of three populations of wolves (figure 1). At t_0, A-wolves may interbreed among them, they are all members of the same species. At t_1, a flood separates one population (A_2) from the other two (A_1). After the flood, A_2 is in fact reproductively isolated from A_1 and, because of the flood, it lives in a very changed environment.[2] After a while, given that they are exposed to different ecological factors (caused by the flood that isolated them), A_2 could start changing, it might be exposed to mutations that are different from those occurring in A_1. Among these mutations, some might affect the capability of A_2-wolves to interbreed with

---

[2] Of course this do not mean that at t_1 a new species has originated: even if A_1-wolves and A_2-wolves cannot interbreed because they cannot physically mate, they *could* interbreed. Mayr's speciation requires something more: A_2 constitutes a new species only whether its members develop mechanisms that prevent them from interbreeding.

the A_1-wolves (perhaps A_1-wolves are no longer attracted by A_2-wolves; or the male gametes of A_2-wolves are no longer able to fertilize the female gametes of A_1-wolves, and so on). Unfortunately, because of the graduality of evolution, mechanisms preventing interbreeding do not arise all at the same time, in all the members of A_2.

Briefly: there is a moment, t_0, in which there is just one species (A); there is another moment, t_2, in which we can say that there are two different species (A_1 and B). But in order to say exactly when the new species B has been originated, it is necessary to choose a conventional time in that vague zone between t_1 and t_2.

3.2. Extinction

With extinction, things seem undoubtedly more definite: who might deny that dinosaurs are extinct? Of course, even concerning extinction, a preliminary convention on what concept of species has to be adopted is necessary, but the weigh of this convention seems to be of little import, at least at a first glance. Maybe because of this (apparent, as we'll see) simplicity, there are very few studies about the nature of extinction and most of us – as well as most biologists – would probably agree with basic definitions along the following lines: extinction is "the end of a species [...]. The moment of extinction is generally considered to be the death of the last individual of that species";[3] or "the end, the loss of existence, the disappearance of a species or the ending of a reproductive lineage" (Delord 2007). But, despite the seeming platitude, I'll argue that, exactly as for speciation, to establish when we may say that a species has become extinct, a convention (in fact more than one) is required.

Consider again, for simplicity, the most popular concept of species, Mayr's concept: a species is a group of interbreeding organisms producing fertile offspring. As for speciation, suppose that biologists reached the agreement that this is the best concept of species for sexual organisms. Mayr's concept seems also quite consistent with the basic definition of extinction reported above. Then suppose also that the extinction of a sexual species is its disappearance, or the ending of the reproductive lineage, and that the criterion for determining whether a species has become extinct is the death of its last member.

---

[3] This definition comes from *Wikipedia*, and I think it is very much like what everyone would say if questioned about the meaning of 'extinction'.

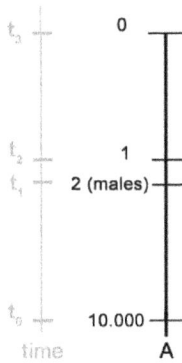

t₃ ... 0

t₂ ... 1
t₁ ... 2 (males)

t₀ ... 10.000

time   A

Figure 2

Imagine that A is a species of wolves (figure 2). At t_0, species A consists of ten hundreds wolves all over the world. (I don't know whether this is a realistic estimate; maybe it is an optimistic one. But consider that, fortunately, in 2010 the Grey Wolf is no longer considered a threatened species according to the IUCN *red list*.) At t_0, A is clearly not extinct. Analogously, at t_3, A has clearly become extinct: no wolves inhabit the world. But, take the slice of time between t_1 and t_3. (i) At t_1, only two wolves are still alive, and they are – say – two males. Should we say that the A is not extinct? Maybe we should say – as several biologists working on conservationist policies do – that it is *functionally* extinct, but this seems an *escamotage* rather than a solution. Or perhaps we should say – as some suggested (see de Quieroz 1999 and Delord 2007) – that a sexual species goes extinct after the disappearance of the last couple, even if individuals of the same sex are still alive. Now (ii) consider A at t_2: just one member is still alive. Is A extinct? More: is A still a species? Finally (iii), from t_2 to t_3, we have the death of the last individual. In order to declare A extinct, should we consider the death or the complete disappearance of its last member? (I know that it could seem I'm splitting hairs, but imagine for instance that, after the death of the last wolf, a laboratory take some genetic material from it – enough to clone it… And this is not science fiction.)

More. Go back to t_1: just two wolves are still alive. Unfortunately, both are males, and wolves cannot switch from sexual to asexual reproduction. Yet we know that several animals may hybridize, either spontaneously or by human intervention, and hybridizing becomes more frequent in critical situations (see Casetta 2009).

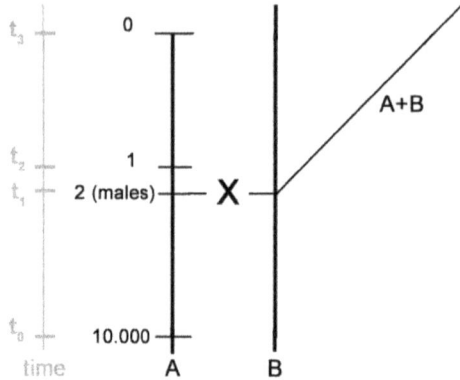

Figure 3

Thus, it is not so implausible to think that two wolves, in a critical situation, might copulate with, say, domestic dogs (species B), producing fertile offspring (hybrids among wolves and dogs are reported – see, for instance, Vilà *et. al.* 2003). In this case, the 'genetic package' of A prosecutes in a new lineage (A + B). At t_3, should we say that A is extinct or rather that it is not, given that it prosecutes in the A + B lineage?

However that may be, what these cases are meant to show is that also in the most standard and apparently clear modality of extinction, decisions have to be taken, and conventions are at work. And it is worth noticing that in every case we have considered, there is no matter of biological fact that could settle the issue.

4 Conclusion

Temporal boundaries of species, namely the point of splitting in the speciation process on the one hand, and the end of the extinction process on the other, are not sharply self-defined: they are, at least in part, a matter of choice, of human conventions. In other words: 'speciation' and 'extinction' are underdetermined concepts because, as we have seen, biological facts alone could not be enough to establish when a new species arises, or when a species goes extinct. Indeed, what is meant by 'speciation' and 'extinction' is firstly dependent on a convention based on a choice of one among the several species concepts available in biology. Moreover, to make sense of speciation a further convention is needed. This convention originated from a choice taken among the several possibilities located in what we have called the 'vague zone' laying between the two extreme poles t_1 and t_2 (figure 1). In a similar way, to make sense of extinction (figure 2) at least three more different conventions are needed, stemming from the relevant decisions to be taken at three different times (recall (i)-(iii)). If this is correct, then the temporal boundaries of species as determined by speciation and extinction – exactly as the temporal boundaries of an organism are determined by its birth and its death – are, at least in part, conventional. Hence, by assumption (a), species are *partly* conventional objects.[4] This means that their identity depends (at least partly) on conventions, i.e., species-directed practices. Had those conventions been different, the individuation and persistence conditions of species would have been different.

Recognizing the role that conventions play in the individuation over time of species could reveal new directions to the debate on metaphysics of species. For instance: a conventionalist approach is already available as an account of ordinary personal identity. According to it, personal identity depends, in some appropriate way, on our conventions. It would be interesting to consider the similarities – if any – between the temporal boundaries of a species and those of a person. Is there any correspondence between the birth of a new species and that of a new person? And between their deaths? Moreover, the issue of identity of species over time and that of the maintenance of biodiversity go hand-in-hand: we cannot expect to safeguard entities that we are not able to identify. Taking into the right account the role played by our conventions in the carving of the natural world could allow us to achieve a novel understanding of biodiversity and of its maintenance, making the role of our conservation policies more active and constructive. Finally, what could the practical consequences of such an approach be? If species boundaries are fixed to a crucial extent by our conventions, what does it mean to *create* new species? What about crossing species boundaries? Think, for instance, of the controversies on the creation of hybrids, transgenics organisms, and chimeras. Such entities have raised – and continue to raise – a heated ethical debate based on the distinction between the organisms and practices that are 'natural' and those that are 'artificial'. But, if conventionalism was right, that distinction ought to be rethought.·

---

[4] I say *partly* because conventions are generally thought to be established on an unconventional *substratum* (a 'stuff' provided by the world) that in turn severally constrains the conventions themselves (see Einheuser 2006).

· I am grateful to Achille Varzi for his suggestions and comments on a previous version of this paper. Thanks are due to the organizers and to the participants of the "Open Problems in the Philosophy of Science" Advanced Training School (Cesena, April 15-17, 2010). In particular I wish to thank Mario Alai, Giovanni Boniolo, Pierluigi Graziani, Valeria Giardino, Andrea Sereni, and Giuliano Torrengo for their helpful comments.

References

Dupré, J. (1993): *The Disorder of Things. Metaphysical Foundations of the Disunity of Science*. Cambridge (MA), Harvard University Press.

Casetta, E. (2009): *La sfida delle chimere*. Milano, Mimesis.

Delord, J. (2007): "The Nature of Extinction". *Studies in History and Philosophy of Biological and Biomedical Sciences*, 38, pp. 656-667.

De Quieroz, K. (1999): "The General Lineage Concept of Species and the Defining Properties of the Species Category", in R.A. Wilson (ed.), *Species: New Interdisciplinary Essays* (pp. 49-89). Cambridge (MA), MIT Press.

Einheuser, I. (2006): "Counterconventional Conditionals". *Philosophical Studies*, 127, pp. 459-482.

Ereshefsky, M. (2001): *The Poverty of the Linnaean Hierarchy: A Philosophical Study of Biological Taxonomy*. Cambridge, Cambridge University Press.

Ghiselin, M.T. (1974): "A Radical Solution to the Species Problem". *Systematic Zoology*, 23, pp. 436-44.

Hull, D.L. (1976): "Are Species Really Individuals?". *Systematic Zoology*, 25, pp. 174-91.

Kitcher, P. (1984): "Species". *Philosophy of Science*, 51, pp. 308-33.

Lewis, D.K. (1969): *Convention. A Philosophical Study*. Cambridge (MA), Harvard University Press.

Maynard Smith, J.; Szathmary E. (1995): *The Major Transitions in Evolution*. New York, W.H. Freeman.

Mayr, E. (1970): *Populations, Species and Evolution*. Cambridge (MA), Harvard University Press.

Schön I.; Martens K.; and Van Dijk P. eds. (2009), *Lost Sex. The Evolutionary Biology of Parthenogenesis*. Berlin-Heidelberg, Springer.

Varzi A. (2011): "Boundaries, Conventions, and Realism", in J.K. Campbell, M. O'Rourke, and M.H. Slater (eds.), *Carving Nature at Its Joints: Natural Kinds in Metaphysics and Science*. Cambridge (MA), MIT Press, pp. 129-153.

Vilà, C. *et. al.* (2003): "Combined Use of Maternal, Paternal and Bi-Parental Genetic Markers for the Identification of Wolf-Dog Hybrids". *Heredity*, 90, pp. 17-24.

Wilkins, J.S. (2009): *Species: A History of the Idea*. Berkeley, University of California Press.

Second Section
Philosophy of Mathematics

Pathways of mathematical cognition

Mario Piazza
University of Chieti-Pescara
m.piazza@unich.it

1 The secret adventures of evidence

In the *New Essays on Human Understanding* (1704), Leibniz argues that even such *self-evident* arithmetical truth as 2 + 2 = 4 can and must be axiomatically proved:

«Definitions:

(1) 2 is 1 and 1
(2) 3 is 2 and 1
(3) 4 is 3 and 1

Axiom: If equals were substituted for equals, the equality remains.
Proof: 2 + 2 = 2 + 1 + 1 (by Def.1) = 3 + 1 (by Def. 2) = 4 (by Def. 3).
Therefore, 2 + 2 = 4 (by the Axiom)» (*Nouveaux essais*, IV, VII § 10)[1].

Leibniz's philosophical purpose is to keep distinct the epistemic notion of self-evidence and that of (mathematical) *truth*: the self-evidence of a mathematical proposition p does not depend upon a proof of p (p may indeed be self-evident even in the absence of a proof), whereas the truth of p is always a proof-relative matter, open to logical evaluation. However, even though self-evidence is assumed not to participate in the production of proofs, if a proof of a certain self-evident proposition turns out be flawed,

---

[1] Leibniz (1981).

then it is plausible to suppose that the culprit is the *pressure* of self-evidence itself, which favoured the overlooking of some deductive steps. That is to say, proving something self-evidential may entail not proving it under ideal conditions: the obviousness allows too much confidence on the *structural* features of the notions it incorporates.

Famously, in the *Foundations of Arithmetic* (1884), Gottlob Frege – who shares with Leibniz the attitude of regarding axiomatisation as an instantiation of an epistemological order – points out a gap in Leibniz's argument: it makes use of the associative law of addition that was not explicitly stated. The emended proof of the proposition $2 + 2 = 4$ is the following: $2 + 2 = 2 + (1 + 1)$ (by def. 1) $= (2 + 1) + 1$ (by associativity) $= 3 + 1$ (by def. 2) $= 4$ (by def. 3)[2]. But as to how the primitive fact of the *general* law of associativity is itself to be know, Frege says nothing: a truth like that is forced on us by Reason itself: the direction of the logical flow – context-invariant – does *coincide* with the direction of the flow of our mathematical awareness. Obviously, the question as to whether logic itself is *justified* lacks of any sense, since it is logic that professionally articulates what justification is.

After Gödel incompleteness theorems, we cannot but dismiss the idea that logical laws mark mathematical (or at least arithmetical) conditions. But we should also dismantle the notion of axiomatisation intended as in the permanent service of a *foundation* – a notion in itself desperately anthropocentric – of mathematical theories. Consider graph theory, for instance: graph theory is a completely *rigorous* albeit nearly unaxiomatised theory, rich of applications to computer science (and, somewhat ironically, to proof theory itself: proofs are acyclic and connected graphs whose nodes are labelled by formulas). Knot theory is another example. Still, the notion of axiomatisation – simply a byproduct of deductive reasoning – is credited with the capacity of telling us something substantive about how mathematics[3].

Traditionally, the axiomatic conception of mathematics involves an appeal to a non-negotiable *evidence* in order to justify our accepting the elite minority of the axioms, and this evidence is supposed to be transferred across deductive inferences from them: as a consequence, a key fact assumed about evidence is that it is of *finite* cognitive depth. Nevertheless, the appeal to some form of primitive evidence has the effect of obfuscating the epistemological dimension of mathematics instead of illuminating it. The general point is that the axioms of a mathematical theory are not the projection of neural conspiracy, but they jointly constitute a kind of

---

[2] Frege (1968).
[3] See Cellucci (1998) and (2008) for an extensive criticism to the "axiomatic ideology".

symptom of an *inferential practice*. Indeed, it is an *inferential practice* that codifies and promotes an economy of primitives, not vice versa. In other words, evidence is *an epiphenomenon of dynamics*: it is *because* we are inclined to make certain inferences among mathematical propositions that we recognize some of these propositions as evident; it is *not* because we recognize some propositions as evident that we are inclined to make inferences from them. We – not axioms for us – *decide* from which conceptual places (non trivial) proofs must be attacked, making the most convenient option with a range of paths in mind at the risk of false or slow starts. These competing paths in turn generate or inspire a spectrum of answers and solutions to unformulated mathematical questions. On the other hand, if the function of a proof were exclusively identified in its capacity of guaranteeing the *truth* of a theorem – as Frege assumes in the *Foundations of Arithmetic* (§2) – then it would be utter mysterious a routinary mathematical activity: giving *new* proofs of *old* theorems[4].

The platitude that mathematics is not its axiomatisations is not a philosophical point, but a mathematical one. Mathematical objects keep their *identity* through a plurality of axiomatic presentations, which thereby represent variations on a common theme. So, the evidential frame has to exist *independently* of any special presentation, being rather the *precondition* of mathematical activity. For example, «the fact that a variety of axiom systems exist for the theory of groups abolishes the presumed privilege of any one system bringing into being the notion of a group, but on the contrary presupposes a pre-axiomatic grasp of the notion of group»[5].

In short, this pre-axiomatic grasp pertains to mathematical epistemology, not to "pre-mathematical" epistemology (if there is one). One might be tempted to say that an axiomatic system cannot provide reasons for ruling out its pre-axiomatic justification without undermining the credentials of its *own* justification.

From the standpoint of logical connectivity, an axiomatic system is simply a way of organizing the logical space, so that "(logical) points" in it are at a *fixed* distance from one another. This distance is measurable by a series of steps of the *same* kind, covering the space without jumps. Like the centralized administration of a country, the axiomatisation of a mathematical theory simply figures as a unitary actor that acquires its authority from the bureaucratic capacity to achieve *coordination* in an intrinsic fashion. Moreover, an axiomatisation is the proper level at which an artificial, *irreversible* time is imposed and controlled, whose aim is to express the notion of *acyclicity*: mathematical "events" – i.e. steps of deduction with a

---

[4] Dawson (2006).
[5] Rota, Sharp, Sokolowski (1988, 382-383).

mathematical content — succeed one another in such a way that no later event causes (i.e. justifies) an earlier one (no loops). Globally, the *evolution* of an axiomatic system is completely determined by its starting points, by its *origin*. Accordingly, axioms are regarded as atomic events, namely as deductions without premises. This means that axiomatic proofs are not adaptive to a changing environment: when a system changes, all its proofs get extinct.

We implicitly divide any proof into past, present and future; at some stage, we can evaluate whether the proof is *finished*. This flux is iconnected, of course, to a genuine epistemic dimension: if the conceptual components of the proof acted *simultaneously*, no cognitive trajectory could be traced and stored (obviously, the fact that we *retain* in our memory the earlier steps of a proof does not constitute part of the reason for the *necessity* of the conclusion, even if *that* memory is a condition for grasping the proof). Accordingly, part of the reliability of a proof arises from its *repeatability* (i.e. not a logical or mathematical notion): if we do not feel convinced by a proof, then we have the right to *re-create* time. This amounts to saying that an axiomatisation is *an ad hoc model of the flow of mathematical information*. There is no extra information to be used for filling the conceptual space, since this space is so arranged to be perfectly homogeneous: the gaps one is expected to plug are inevitably "obvious". The information flow is *ad hoc* not in the sense that the package of axioms represents an intellectual wager, but because its selection is generally encouraged, authorized or *dictated* by the theorems themselves: axioms are *logically* more powerful than theorems, but theorems are *cognitively* more powerful than axioms. What emerges is that an adequate understanding of mathematical knowledge requires us to unravel the tension and the resulting mediation between the *explicitness* of deduction (read: logic) and the *implicitness* of cognition (read: what is outside the realm of logic, essentially).

2 Proofs *vs.* logic

At the beginning of the twentieth century, Poincaré warned against the tendency to reduce a mathematical proof to a network of logical inferences:

> Should a naturalist who had never studied the elephant except by means of the microscope think himself sufficiently acquainted with that animal? Well, there is something analogous to this in mathematics. The logician cuts up, so to speak, each demonstration into a very great number of elementary operations; when we have examined these one after the other and ascertained that each is correct, are we to think we have grasped the real meaning of the demonstration? Shall we

have understood it even when, by an effort of memory, we have been able to repeat this proof by reproducing all these elementary operations in just the order in which the inventor had arranged them? Evidently not; we shall not yet possess the entire reality; that I know not what, which makes the unity of the demonstration, will completely elude use[6].

This means that, although the relation of "dynamic equilibrium" holding among the components of a proof is shaped by what is *logically possible*, on an epistemological level, logic is constitutively incapable of giving us, so to speak, the *spirit* of the proof, i.e. what is that binds its elements together to make it *one* thing. In other words, the fact that no mathematical proof is possible without being logically decomposable does not make the *actual* decomposition of a proof *the* reason for the authentic understanding of its textured map, i.e. its full scenario. Rather logic encourages the fragmentation of what can ultimately be understood only as a unitary phenomenon in virtue of the cohesive forces of its parts. The "cutting up" – the logical divide-and-conquer – destroys what the mathematical knower seeks to understand. Poincaré's essential point, then, is that logic recognizes as its own the rules of inference that are correctly applied in the course of a proof, but it remains by nature silent on which rules and protocols are to be selected since it is unable to tell *why* a proof must have a certain shape or architectural configuration. That is the reason why the *logical* acceptance of concrete inferences among mathematical concepts – licensed by abstract rules – is not on a par with their *mathematical* acceptance.

From proof theory we have learnt that the phenomenon primarily responsible for the *dynamic* in a proof is the cut-rule – a generalisation of the well-known rule of *modus ponens*: from the premises A and $A \rightarrow B$, conclude B – which corresponds to the ubiquitous mathematical tactics of using intermediate lemmas or general theorems within a proof[7]. A *normalisation theorem* provides, then, an effective procedure for eliminating any application of the cut-rule from the proofs of a logical calculus, making *explicit* – but at the cost of a distortion of the original logical path – the really *useful* information content of lemmas. That is, to every proof with cuts is associable a proof without cuts, called *normal form*, which is an explicit and *combinatorial* version of the original proof. From this general perspective, then, the meaning of normalisation lies exactly in the fact that computation can be controlled by logic, where computation is viewed as the logical flow of information within a proof. However, we can make proofs without lemmas – and it is a very surprising result – but this entails an epistemic loss, that is a loss of understandability since the resulting proofs

---

[6] Poincaré (1946, 217).
[7] Gentzen (1935).

are much larger, complex and artificial than the original ones (consequently we have also a loss of control of the mathematical time). In this sense, then, logic teaches us why we do *not* understand proofs: there are true statements which take too much time to be understood and there are statements that can be understood only through the use of the cut-rule. Thus, the cut-rule is situated exactly in the middle between *simultaneity* (conclude *B*, without *before A* and *A → B*) and a *decompression* of mathematical time in which "events" succeed one another in such a way that each earlier event occurs in a later one.

3 Conclusion

Mathematics actually appears as a "three dimensional manifold" – borrowing an image by Giuseppe Longo – that is to say the product of the interplay of the logical, the formal and the geometrical, with distinctive but not separate cognitive roles[8]. The essays in this volume by Valeria Giardino, Gabriele Pulcini, Andrea Sereni and Gianluca Ustori speak in favour of this interplay.

Proofs cannot be artificially limited to a single dimension since they express – without any occurrence of informational conflicts – a stratification of levels: mechanical (read: formal), constructive, geometrical. The idea is that the essentialistic question: "What is a proof?" – which ultimately invites a (trivial) logical answer – should be replaced by the more salient question: "What means to understand a proof?". Poincaré's answer was that the grasp of a proof coincides with the perception of its unity (via a cognitive, not logical, mechanism). Thus, understanding a proof is essentially equivalent of delimiting its particular *shape* without decomposing the whole. That delimitation – which "contours" the integrity of proof – entails via intuition the capacity of compressing the mathematical information from which the proof is constructed, as we have seen. On the other hand, logic leaves out the mathematical *project* of the proof, being unable to decide on its own initiative *when* and *which* applications of the rules of inferences must be activited or inhibited. Certainly, *directionality* is a necessary feature of deduction: but the directionality of a proof is not the sum of the directionalities of all the deductions involved.

However, although the experience of perceiving the unity of a given proof makes no explicit allusion to other instances of proofs, it is possible to *individuate* proofs only by thinking of them as *members* of a category, and so logic is obviously immanent in this individuation. One can put the point

---

[8] Longo (2005).

in this way: it is immediately when we encounter a proof that we have never seen before that logic intervenes. Furthermore, two proofs can be *equivalent* even though their boundaries are *different* (i.e. the two proofs may have different paths leading to the same result): in that familiar case, it is not appropriate to assume that intuition equips us with the sense of equivalence, since information about the aspects in which boundaries of proofs are dissimilar is not relevant here. Rather, it is logic that makes available clues essential to the discrimination of the equivalence relations holding between proofs. Logic can be characterized *as a mechanism that helps us state and remember similarities* (recurrences). In sum, logic looks at those characteristics *shared* by proofs, namely it deals with the *whole space* of proofs. So, the unity of *proofs* – not of *an* actual proof – is a question of logic. And it is a highly sophisticated one: proof theory advertises itself as the area of logic that studies the general structures of mathematical proofs and the character of the relationships proofs bear to each other.

Since proofs live in an interactional environment, one may say that we understand a proof π when we are able to embedd π in the space of all proofs. Since π is a *part* of this space, any adequate understanding of π entails focusing on the space which reveals the relation of π to *others* proofs; and this sense primarily and more deeply emerges throught the *interaction* of π with others proofs. Consequently, being able to understand a proof means being able to make interact its conclusion with some conclusions of other proofs. In other words, *the notion of understanding itself is dominated by that of interaction,* so that the interaction among proofs is *cognitively* more powerful than the proofs themselves.

References

Cellucci, C. (1998): *Le ragioni della logica*. Roma-Bari, Laterza.

Cellucci, C. (2008): "Why proof? What is a proof?" in G. Corsi and R. Lupacchini (eds.), *Deductions, Computation, Experiment. Exploring the Effectiveness of Proof*, Berlin, Springer-Verlag, pp.1-27.

Dawson, J.W. (2006): "Why do Mathematicians Re-Prove Theorems?". *Philosophia Mathematica* (III) 14, pp. 269–286.

Frege, G. (1968): *Foundations of Arithmetic*. Oxford, Blackwell.

Gentzen, G. (1935): "Untersuchungen über das logische Schliessen". *Mathematische Zeitschrift*. 39: pp. 176-210: 405-431; [English translation in M. E. Szabo (ed.) *The Collected Papers of Gerhard Gentzen*. Amsterdam, North Holland, 1969, pp. 132-213].

Leibniz, G.W. (1981): *New Essays on Human Understanding*. Trans. and ed. P. Remnant and J. Bennett. Cambridge, Cambridge University Press.

Longo, G. (2005): "The Reasonable Effectiveness of Mathematics and its Cognitive Roots", in L. Boi (ed.), *New Interactions of Mathematics with Natural Sciences*. Singapore, World Scientific.

Poincaré, H. (1946): "The Value of Science", in G. Halsted (ed. and trans.), *The Foundations of Science*. Lancaster, PA, The Science Press.

Rota, G.-C.; Sharp, D. H.; Sokolowski, R. (1988): "Syntax, Semantics, and the Problem of the Identity of Mathematical Objects". *Philosophy of Science*, 55, pp. 376-386.

The 'Extraction' of the Principles
of Construction in Number Theory
Some ideas for a research program

Gabriele Pulcini

University of Chieti-Pescara

gab.pulcini@gmail.com

1 Principles of Proof and Principles of Construction

During the last few years, Giuseppe Longo has refreshed the debate on the foundations of mathematics by proposing a new epistemological paradigm based on the cleavage between *principles of proof* and *principles of construction*.[1] The table below displays the main features of such a paradigm through a list of opposite epistemological concepts.

principles of proof	*principles of construction*
formal justification	Constructions
rewriting systems	interaction between math. Structures
encoding	'geometry'
absolute 'newtonian' time	relative temporalities
mathematical induction	prototypical proofs, ...

On the one hand, the principles of proof concern the part of mathematics which admits a treatment in terms of a formal rewriting

---

[1] Longo (2006).

system, i.e. a finite set of axioms together with a finite set of rewriting rules. A mathematical discipline completely placed within the ambit of the principles of proof turns out to be faithfully encodable *à la* Gödel and so mechanically reproducible without any loss of information. Mathematical induction constitutes the principal and most powerful demonstrative method concerning this ambit. The inductive reasoning links the principles of proof to standard achievements of computability theory, essentially based on recursive algorithms, and so it imposes a strictly linear and absolute ('newtonian', we could say) temporality to proofs and computations.[2]

On the other hand, the field of the principles of construction is more vague, but it can be quite precisely defined in a negative way by saying that it gathers all the mathematical behaviours escaping from the ambit of the principles of proof and, in particular, the 'encoding sensible' structures. These kinds of concepts are called 'geometrical' in a sense we will try to specify later. In the same way, the demonstrative methods strictly pertaining to the principles of construction are those not referable to any form of mathematical induction. This latter aspect opens to nonstandard temporalities and computations essentially based on concurrency and interaction between mathematical structures.

In this regard, the incompleteness results — the most celebrated ones by Gödel,[3] but also the so-called 'concrete' incompleteness phenomena[4] — would show that the conceptual richness of arithmetic cannot by enclosed within the limited ambit of the principles of proof.

The just explained dichotomy may recall some other well-known approaches, for instance the classical paradigm based on the distinction between intensional and extensional concepts or the more recent one introduced by Gauthier and consisting in the opposition internal / external logic.[5] Nevertheless, by stressing the crucial notion of 'geometry', Longo's proposal offers the advantage of an explicit link to both computability theory and cognitive science.

In these few pages, we aim to address the problem of the 'extraction' of the principles of construction specifically concerning number theory. In particular, we are interested in singling out the demonstrative methods which logically diverge from mathematical induction. In Longo (2000), Longo has indicated the proofs based on prototypical arguments as an example of deductive configurations alternative to those based on inductive strategies. The example provided by Longo consists in the well-known symmetry stressed for proving that the sum of the first n naturals equals

---

[2] Smorynski (1991).
[3] Gödel (1931).
[4] Harrington and Paris (1978).
[5] Gauthier (2002).

$n(n+1) / 2$. Consider a generic n, then write a sort of $n \times 2$ matrix in which the first row is filled by the series 1, ..., n and the second row is simply obtained by reversing the first one. Now, the sum performed along each column always produces $n + 1$ therefore the claim of our theorem follows straightforwardly.

1	2	...	n
N	n - 1	...	1
n + 1	n + 1	...	n + 1

Leaving aside our personal doubts about the effective justificative power of the specific example just displayed (is there an hidden inductive mechanism at work behind the notation '...' ?), prototypical strategies seem unable to face the intrinsic complexity and difficulties associated with most of the actually interesting problems arising in number theory. The history of this discipline suggests we consider the fermatian principle of the indefinite descent as, pratically speaking, the unique method whose demonstrative power really deserves a comparison with mathematical induction.[6] As we will show, an in-depth logical investigation concerning the demonstrative mechanism of the indefinite descent may also provide an interesting notion of 'geometry' for number theory essentially based on the concept of numerical form.

2 Indefinite Descent

2.1. The method

The first use of the descent goes back to the euclidean Elements (Book VII, prop. 31). Nevertheless, really meaningful applications of this method were firstly proposed by Fermat in his work on Diophantine analysis.[7] After the French mathematician, the great number theorists Euler and Lagrange, inspired by the fermatian work, proposed important applications of the descent. We mention from Bussotti (2006) some of the most important theorems proved by descent:

---

[6] Ore (1998); Weil (1984).
[7] de Fermat (1899).

any natural number turns out to be the sum of 3 triangular numbers (Fermat 1601-1665),

there is no Pytagorean triangle whose area equals a square number (Fermat),

for any triple $x, y, z \in \mathbf{N}$, $x^3 + y^3 \neq z^3$ (Euler 1707-1783),

any prime number having the $4n + 1$ form is the sum of 2 squares (Euler),

any natural number can be represented as the sum of 4 integer squares (Lagrange 1736-1813).

The general shape of the fermatian technique can be summarised as follows.

Indefinite descent. We aim to prove that, for any $n \in \mathbf{N}$, the property $P(n)$ holds true. We proceed by contradiction: suppose that there exists a $k \in \mathbf{N}$ such that $\neg P(k)$. We show that $\neg P(k)$ implies the existence of a $k' \in \mathbf{N}$ such that $\neg P(k)$ and $k' < k$. But this would mean that the finite segment $[0, k]$ 'contains' more than k naturals which is, by the structure of $\mathbf{N}$ itself, absurd.

We report from Jones and Jones (2006) a very simple proof by descent.

Example. We aim to show that there is no Pythagorean triple (a, b, c) with $a = b$; in other words: the equation $2x^2 = y^2$ has no integer solutions. The proof is by contradiction. If such a triple (a, a, c) exists, then $c^2 = 2a^2$, so c^2 is even and hence so is c. Putting $c = 2c'$ we get $4c'^2 = 2a^2$ and so $a^2 = 2c'^2$, showing that a^2 is even and hence so is a. Putting $a = 2a'$ we see that $c'^2 = 2a'^2$, which gives another isosceles Pythagorean triple (a', a', c') with strictly smaller terms than the first one. Applying this process again to our new triple, we can get a third triple (a'', a'', c'') with yet smaller terms. By repeating the process, we get an infinite sequence of such triples. Their first entries then form a strictly decreasing infinite sequence $a > a' > a'' > \ldots$ of positive integers (see **Figure 1**), which is impossible: any such sequence of integers must sooner or later contain negative terms. Thus, there can be no isosceles Pythagorean triple.

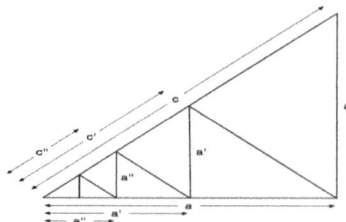

Figure 1

2.2. Infinite descent *vs* mathematical induction

As far as the relation between mathematical induction and indefinite descent is concerned, logicians and number theorists are usually in great disagreement: whereas the majority of logicians assert the equivalence of these two methods, number theorists consider them as two really different demonstrative techniques.

From a strictly formal point of view, logicians uphold a triviality: we can easily pass from the standard formal rendition of mathematical induction (1) to the so-called complete induction (2) and then to point (3), representing the formal indefinite descent (3):[8]

(1) $(P(0) \wedge \forall x \, (P(x) \rightarrow P(x + 1))) \rightarrow \forall x P(x)$

(2) $\forall x \, (\forall y \, ((y < x) \rightarrow P(y)) \rightarrow P(x)) \rightarrow \forall x P(x)$

(3) $\forall x \, (\neg P(x) \rightarrow \exists y \, ((y < x) \wedge \neg P(y))) \rightarrow \forall x P(x)$

More specifically, logicians usually maintain that the descent mechanism would be nothing but a sort of reverse complete induction stressing the following implication:

$\neg P(n) \rightarrow (\neg P(n-1) \vee \ldots \vee \neg P(0))$.

In this case, we would have a sort of descent proceeding, as the fermatian method does, by performing irregular 'jumps'. Nevertheless, it is sufficient to remark that the specific descent triggered by the 'reversed' complete induction will consist in a *finite* number of steps leading to a *local* contradiction which is always of the form $\neg P(0) \wedge P(0)$. Though not explicitly mentioned in point (2), the basis is however present as a key point in any inductive argument (besides, points (1) and (2) are two equivalent formulations!). From this point of view, the indefinite descent radically diverges from induction. As a matter of fact, an argument by descent is based on the possibility of *indefinitely* lengthening the descent so as to achieve a *structural* contradiction, i.e. a contradiction against the structure of **N** itself which seems to be closer to a violation of the pigeonhole principle.

The framework of the first order formal arithmetic seems not able to grasp such a cleavage, therefore we propose to place the indefinite descent in the area of the principles of construction, beside prototypical strategies.

---

[8] Gauthier (2009).

3 A (strictly) constructive approach

In Gauthier (2009), the problem of the comparison between induction and indefinite descent has been already addressed from the intuitionistic point of view. Nevertheless, we propose to perform a further step in this direction by considering the same problem from a strictly *procedural* point of view. According to this setting, it would be possible to assert the equivalence at issue only in presence of an *effective procedure* able to turn a proof by induction into another one by descent and/or *vice versa*. The history of number theory gives evidence in favour of the great difficulty of establishing such a procedure for non trivial proofs, so: how could we mathematically prove their effective divergence? A possible way to achieve such a result could be that of proving by descent a mathematical proposition shown to be independent from the so-called Peano Arithmetic.[9]

There are often computational strategies associated with proofs by indefinite descent, the so-called *strategies by reduction*.[10] So, a divergence between induction and descent might indirectly imply a divergence between classical recursive procedures and computations by reduction. As already seen, the indefinite descent often proceeds by performing 'jumps' upon **N** which are, at the same time, *non-deterministic* and *regular*. On the one hand, they are non-deterministic, because we are not able to predict, at each step, how much the magnitude will decrease (notice that the example here proposed is too simple to be meaningful in this sense). On the other hand, all the quantities touched during the descent share a precise *numerical form* and the main difficulty in providing a proof by descent consists indeed in singling out such a form (in our example the form at issue is the very simple one denoting even numbers). From this point of view, the descent mechanism seems to underline a sort of 'quantum behaviour' in number theory which, in our opinion, would deserve to be investigated, priorly, in a comparison involving the standard achievements of quantum computation.

In conclusion, we remark that the just mentioned concept of numerical form, which seems to be the core of the descent demonstrative technique, may provide an interesting insight in order to elucidate the notion of 'geometry' in number theory.

---

[9] Harrington and Paris (1978).
[10] Bussotti (2006).

References

Bailly, F. ; Longo, G. (2006): *Mathématiques et sciences de la nature*. Paris, Hermann.

Bussotti, P. (2006): *From Fermat to Gauss*. Augsburg, Dr. Erwin Rauner Verlag.

de Fermat, P. (1899): *Œuvres*. Volume II. Paris, Gauthier-Villars.

Gauthier, Y. (2002): *Internal Logic*. Synthese Library, Dordrecht, Kluwer.

Gauthier, Y. (2006): "La descente infinie, l'induction transfinie et le tiers exclu". *Dialogue* 48, pp. 1-17.

Gödel, K. (1931): "Über formal unentscheidbare Sätze der Principia Mathematica und verwandter Systeme I". *Monatshefte für Mathematik und Physik*, 38, pp. 173-198.

Harrington, L.; Paris, J. (1978): "A mathematical Incompleteness in Peano Arithmetic", in J. Barwise (ed.), *Handbook of Mathematical Logic*. Amsterdam, New York, London, North-Holland Publishing Company.

Jones, G.A.; Jones, J.M. (2006): *Elementary Number Theory*. London, Springer Verlag.

Longo, G. (2000): *"Reflections on Incompleteness, or on the Proofs of Some Formally Unprovable Propositions and Prototype Proofs in Type Theory"*, in P. Callaghan et al. (eds), *Types for Proofs and Programs, Durham, (GB)*. *Lecture Notes in Computer Science*, vol 2277. Springer, pp. 160 - 180.

Ore, O. (1988): *Number Theory and its History*. New York, Dover.

Smorynski, C. (1991): *Logical Number Theory I*. Berlin, Springer-Verlag.

Weil, A (1984): *Number Theory*. Boston, Birkhäuser.

The Constitution of Mathematical Objects
A Critique of Gödel

Gianluca Ustori
University of Siena
gianlucaustori@hotmail.com

1 Gödel's philosophy of mathematics

Kurt Gödel (1906-1978) propounds a philosophy of mathematics centered in a strong mathematical realism about concepts and sets, a conviction that he held from 1925.[1] In the Russell paper (1944) he writes that

> classes and concept may [...] be conceived as real objects, namely classes as 'plurality of things' or as structures consisting of a plurality of things and concepts as the properties and relations of things existing independently of our definitions and constructions.[2]

Gödel confers an ideal existence not only to classes, as in the extensional approach, but also to concepts, intended as intensional entities, and searches without success for a theory of concepts solving intensional paradoxes just as simple type theory or axiomatic set theory solves extensional paradoxes.

Some years later, in the Gibb's Lecture (1951), he links Platonism with a strong conception of intuition or perception: «concepts form an objective

---

[1] The Grandjean questionnaire, CW IV (447).
[2] Gödel (1944, 128).

reality of their own, which we cannot create or change, but only perceive and describe».[3] Concluding the Lecture, he remarks that

the Platonistic view is the only one tenable. Thereby I mean the view that mathematics describes a non-sensual reality, which exists independently both of the acts and [of] the dispositions of the human mind and is only perceived, and probably perceived very incompletely, by the human mind.[4]

Gödel opposes this strong kind of intuition to space-time Kantian intuition, which is limited to the sensual and finite dimension.

Hilbert's intuition is of this kind and is not sufficient to prove the consistency of arithmetic, which was the aim of Hilbert's program: in fact, the incompleteness theorems revealed its failure. Gödel observes that «our intuition tells us the truth not only 7 plus 5 being 12 but also [that] there are infinitely many prime numbers and [that] arithmetic is consistent. How could Kantian intuition be all?»[5] Discussing Cantor's continuum hypothesis, to which Gödel confers a determinate truth value (he believes that it is false) despite its undecidability from the axioms of set theory, Gödel also criticizes Brouwer's intuition, because intuitionism «denies that the concepts and the axioms of classical set theory have any meaning (or any well-defined meaning)»[6] and is therefore «destructive in its results».[7] In 1964 Gödel repeats, on the contrary, that «despite their remoteness from sense experience, we do have something like a perception also of the objects of set theory, as is seen from the fact that the axioms force themselves upon us as being true».[8]

A similar kind of existence pertains, in Gödel's opinion, to physical objects and to mathematical concepts. In the Russell paper he affirms that

the assumption of such objects [classes and concepts] is quite as legitimate as the assumption of physical bodies and there is quite as much reason to believe in their existence. They are in the same sense necessary to obtain a satisfactory system of mathematics as physical bodies are necessary to obtain a satisfactory theory of out sense perceptions.[9]

---

[3] Gödel (*1951, 320).
[4] Gödel (*1951, 323).
[5] Wang (1996, 217).
[6] Gödel (1947, 181).
[7] Gödel (1947, 179).
[8] Gödel (1964, 268).
[9] Gödel (1944, 128).

In the 60's Gödel remarks that «the question of the objective existence of the objects of mathematical intuition [...] is an exact replica of the question of the objective existence of the outer world».[10] There is therefore a strong analogy, in their mind-independent existence, between mathematical and physical reality, and, as acts of knowledge, between mathematical intuition and sense perception, by which we reach respectively the material and the ideal world.

Gödel studied Leibniz between 1943 and 1946[11] and developed the ideal of metaphysics as a «monadology with a central monad [namely, God]».[12] He also studied Kant, and, as Wang tells, «had to take into account Kant's criticism of Leibniz»[13] and in general of every realist conception of science as knowledge of the things in themselves. Gödel comes to the conviction that a foundation of realism can only be given along renewed and corrected Kantian lines, i.e. along a sort of transcendentalism explaining the link between mind and reality, in particular between mind and the «objective reality of concepts and their relations».[14]

In 1959 Gödel discovers Husserl's phenomenology and takes it as a philosophical frame in which he thinks to find a justification for his conceptions about Platonism and intuition. In the words of Wang, Gödel «saw Husserl's method as promising a way to meet Kant's objections»,[15] and we can pick out two of Husserl's ideas that impressed him: Husserl's conception of *eidetic intuition*[16], i.e. of intuition of essences (that are nearly the same as Gödel's concepts[17]) which is what Gödel needed for his strong (non-Kantian) mathematical intuition, and Husserl's conception of a *universal and transcendental correlation between consciousness and reality,*[18] which is what Gödel needed to bridge the gap of knowledge between mind and world, in particular to solve what is nowadays called the "problem of access" to mathematical entities. Gödel believes that phenomenology clarifies, corrects and develops Kant's thought, because it «avoids both the death-defying leaps of [German] idealism into a new metaphysics as well as the positivistic rejection of all metaphysics».[19]

---

[10] Gödel (1964, 268).

[11] Wang (1996, 7).

[12] Wang (1996, 8).

[13] Wang (1996, 8).

[14] Gödel's letter to Schilpp, february 3, 1959, CW V (244).

[15] Wang (1996, 8).

[16] Husserl (IDEEN, ch. 1) and (LU II, Second Investigation).

[17] Wang (1996, 167).

[18] The discovery of this correlation is at the origin of phenomenology took place during work on the Logical investigations around 1898 (Husserl (KRISIS note to § 48), but only from 1905 on it was inserted in a transcendental framework (Husserl (IP, Lecture 1).

[19] Gödel (*1961/?, 387).

According to Gödel, Kant is right in the discovery of the transcendental link between mind and world, but has to be corrected in «his conviction of the unknowability [...] of the things in themselves».[20] Gödel has perhaps seen Husserl's motto "back to the things in themselves!"[21] as a sign of his overtaking of Kantian conviction.

In the light of Gödel's claims, it is important to clarify the constitution of mathematical objects and the structure of mathematical ontology, as it is presented by transcendental phenomenology, so that we can judge if Gödel's ideas about realism and intuition about mathematics and physics are defendable and if his understanding of Husserl is deep or rather misleading.

2 The constitution of mathematical objects

According to transcendental phenomenology, every object is the result of a process of constitution taking place in our consciousness and rooted in our life-world.

Descriptive phenomenology, presented by Husserl in the *Logical investigations* (1900-01), starts with a *reflection* on our inner experiences (*Erlebnisse*) or acts of consciousness, which exhibit a structural feature, intentionality, that is the directedness of every act to an object. *Transcendental* phenomenology, sketched by Husserl in *The idea of phenomenology* (1907) and then systematically presented in the *Ideas* (1913), starts when, with the phenomenological *reduction*, or *epochè*, we suspend the general and basic belief in the existence of the world and in the validity of the truths about it, given by common sense as well as by sciences, and so doing we pass from the *natural attitude* to the *phenomenological* one. After the reduction, conscious acts reveal the *hyle-noesis-noema* structure: the *hyle* is the residue of sense data, deprived of every explanation or theory; the *noesis* is the intentional structure of the subjective act, that animates the hyle, giving it a sense; the *noema* is the objective result and correlate of the sense-giving activity of the noesis, and its objectivity is owed to its invariance compared to the continuous variation of inner experiences. Intentionality is now described as a noetic-noematic structure, and the directedness is that of the noesis to the noema. This one is a projection (in the Latin sense of "throwing-forward") of consciousness, and it is what we, in the natural attitude, take as the obviously existing

---

[20] Gödel (*1946/49-B2, 244).
[21] This motto has an origin in the sentence «We want to go back to the things in themselves» (LI II, Introduction § 2).

object,[22] which, albeit normally intended as transcendent, appears to phenomenological analysis as transcendentally constituted.

In his later years, Husserl realizes that the noetico-noematic constitution is rooted in the pragmatic dimension of human action in our surrounding life-world.[23] The dimension of praxis is at the origin of human consciousness itself as distinguished from that of the other animals, as is suggested by the anthropologist Arnold Gehlen,[24] and the "I", before being an "I think" (noetic sphere), is an "I can"[25] (practical sphere) with its complex of abilities centered in the movement of the lived body and in the use of the hands in connection with the senses. Luis Romàn Rabanaque observes that «the practical correlates of these acts should perhaps be called *practical noemata* or *pragmata*».[26] Objects are therefore constituted first as pragmata (objects of action) and only then as noemata (objects of intellection).

Coming to mathematical entities, Enrico Giusti analyses Euclide's definitions of straight line, circle and sphere, and interprets them as syntheses of the practical operations of land-surveying,[27] concluding that «mathematical objects come [...] from a process of objectualization of procedures».[28] Similar considerations are made by Husserl in *The origin of geometry* (1936):[29] as perceptual objects, also mathematical ones are constituted in an interplay between observation and action. But what about complex numbers or abstract groups? Giusti suggests that in these cases we have some more abstract operations, for example about the resolution of algebraic equations, but the process is the same: mathematical objects

> come in firstly as research tools, proof methods originated from innovative ideas; secondly they become both solutions of problems and objects of study, and at the end of this process they acquire an out-and-out objective existence.[30]

From procedures, which in years, decades or centuries are stabilized for their efficacy, new unifying concepts emerge, but this process can give no warranty for the future stability of such concepts and procedures:

---

[22]The word "object" also comes from the Latin root "thrown against", that characterizes it as the result of an act, which can be an act of consciousness or, as we will see, a practical action of the body.

[23] Husserl (KRISIS).

[24] Gehlen (1940).

[25] Husserl, (IDEEN II, § 59).

[26] Rabanaque (2005, 457).

[27] Giusti (1999, 24).

[28] Giusti (1999, 26, my translation).

[29] Husserl (KRISIS, Appendix III).

[30] Giusti (1999, 32, my translation).

constitution is always fallible, and in the history of mathematics we find some concepts, like indivisibles and infinitesimals, which had a very short existence. Giusti concludes that

> the best answer to the question if mathematical objects are invented or discovered is that they are both invented and discovered. We don't have to believe that they are invented and discovered at the same time; on the contrary, they are first invented as proof procedures, and then discovered as mathematical objects. The realism of the major part of mathematicians comes from here: they are inclined to speak of invention about proof techniques, but maintain that the objects they deal with have an objective reality, i.e. that they exist before and independently of their discovery.[31]

Here Giusti distinguishes the two levels of constitution-invention and intuition-discovery, similarly as in phenomenology, where we have *no intuition without constitution*, and where what appears to be independently and eternally existing is, as a deeper analysis shows, the result of a concealed constitution process.

Our world is composed of a lot of different kinds of objects. We have material things, but also states of affairs, concepts, collections, propositions, theories, etc. An important phenomenological distinction is that between *perceptual* and *categorial objects*. The second ones are higher objects, distinguished on one side in *essences* and on the other side in *synthetic categorial objects*, and these are often indicated simply as categorial objects tout-court. While perceptual objects are constituted and intuited in the act of *perception*, to reach essences we need *eidetic intuition*, and to reach categorial objects we need *categorial intuition.*[32]

We can see a tree and we can see *that* the tree is higher than the car. *To see* is obviously not the same as *to see that*: perception gives us things as the tree and the car, but not states of affairs as "the tree is higher than the car". In a similar way, perception gives us the tree and the car, but not the set composed by "the tree and the car" as a unity. States of affairs and sets are two kinds of categorial objects, founded on perceptual objects, but irreducible to them.

Starting with a categorial object, we can accomplish another act, *formalization*,[33] that deprives every perceptual component of the categorial object of its sensible-material content (hyle), so that we obtain an empty structure, called formal categorial object. From "the tree is higher than the

---

[31] Giusti (1999, 75, my translation).
[32] Husserl (LU II, Sixth investigation, ch. VI).
[33] Husserl (LU II, Sixth investigation VIII), (IDEEN First section ch. 1), (FTL First section ch. A1).

car" we obtain "t > c", or "t R c", where R is an order relation. Formalization is linked to the use of variables, and already existed in Greece, pertaining to logic; but for mathematics we had to wait until the XVIIth century, when Descartes introduced it in algebra. Husserl sees the development of mathematics in the XIXth century as characterized by the discovery of the formal nature of mathematics. From arithmetic as science of quantity, geometry as science of space, and logic as science of correct reasoning, we arrive at a new image of these disciplines, seen as sciences of formal structures of different kinds, but without a concrete subject in the real world.

We firstly characterize a *mathematical object* as a *formal categorical object*, i.e. as a formal structure, and we can obtain an endlessly growing hierarchy of such objects.

With eidetic intuition or *generalization*,[34] then, we constitute from a hyle an eidos or essence or concept, instead of a perceptual object. To a tree, the eidos of tree corresponds; to the number 3, that of natural number; to the feature of the tree "being higher" that the car, the essence of order; and so on. Husserl underlines that eidetic intuition is not an empirical kind of induction: we *see* the eidos, we do not need a lot of examples to abstract from. To clarify our intuition of an essence we can use a systematic method, free variation in imagination: starting with an example, we vary our essence with our fantasy, until we reach its structural limits. The eidos of tree is subordinated to the eidos of a living being, and that of a triangle to that of a plane figure. We must not confuse generalization and formalization: the first one gives formal objects and structures,[35] the second one essences or concepts, that can be material ("tree") or formal ("natural number").

In the realm of essences we distinguish morphological or vague from exact ones: "sweet" is vague, "triangle" is exact. Mathematics is an eidetic and exact science, and deals only with *exact essences*, that are obtained with the act of *idealization*.[36] This act is particularly important for the genesis of geometry (mathematics of the continuum), where we deal with indivisible points, lines without thickness, perfect circles, and so on. Modern physics is a geometrization of space-time, and for this reason idealization is the

---

[34] Husserl (LU II, Second investigation), (IDEEN First section ch. 1).

[35] The number 3 is a formal object, and "natural number" is a formal concept: in this terminology "object" means "individual" and "concept" means "universal". Another choice is calling both "concepts", if we confine the word "object" to the material individuals, and if we use "concept" for both the results of formalization and of generalization. In this sense the number 3 would be a formal individual concept, and "natural number" would be a formal general concept. I prefer the first terminology (because with the other we do not distinguish clearly between formalization and generalization) but Longo, for example, uses the second one in the title of Longo (2007).

[36] Husserl (IDEEN First section, ch. 1), (FTL Second section ch. 3), (KRISIS § 9).

constituting act of the exact ontology of physics. Idealization also plays an important role in arithmetic (mathematics of the discrete), constituting *infinity* with the "and so on" structure and the connected principle of mathematical induction. Examples of exact and infinite essences are not to be found in our world: we can find a tree, but not a (perfect) circle or an (infinite) straight line. They are "perceptively false" essences,[37] but nevertheless important to unify "from a distance", in a panoramic way, a lot of perceptual things, and that would be impossible with a "perceptively true" morphological essence. Longo writes that

> mathematical objects are limit constructions, obtained by a conceptual 'critical' transition, where the constitutive contingency is lost at the limit. Euclid's line with no thickness or the 'transcendental' number π is the result of a geometric construction, pushed to the limit. But, in the end, their objectivity does not depend on the specific-contingent and more or less abstract reference to actual traits or sequence of rational numbers, needed to conceive or present them: at the limit, the transition to infinity provides us with a perfectly stable conceptual object.[38]

Exact and infinite objects, according to phenomenology, are accepted as "perfectly stable" invariants of our experience, constituted through a limit process: after the critical transition we have the new object, the constituting path is lost, and the object appears as mind-independently existing. Husserl explains, against Platonistic thinkers like Galileo and Gödel, that the physical world should not be seen as the "real" world of things beyond our "illusory" perceptual appearances, but as an exact structure, constituted through idealization from the vagueness of the life-world.

A last consideration has to be devoted to the *temporality* of objects, and particularly of mathematical ones. Mathematical objects and truths are usually taken as timeless, and Husserl also distinguishes the determinate time position of a concrete state of affairs from the absence of such a position in the case of the formal state of affairs "2+2=4". The late Husserl, however, clarifies that timelessness does not mean existence in a timeless world, but on the contrary it has to be intended as *omni-temporality*,[39] i.e. constituability in every possible time. Because of their formal character, mathematical structures are constituable without material or temporal constraints: they are not to be taken as existing in a «second reality»[40]

---

[37] Tragesser (1991).

[38] Bailly - Longo (2008, 249).

[39] Husserl (FTL Second section § 58), (CM Fifth meditation § 55), (EU Second section § 64).

[40] Gödel (*1953/9-III, 353).

without matter and time, but as formal structures that are constituable from this world from all possible matter and at every possible time.

It emerges from this phenomenological analysis of mathematical objects that mathematical intuition and knowledge "discover" only what was previously constituted and constructed. Phenomenology admits no mind-independent reality, and reveals that seemingly transcendent objects and concepts are constituted noemata: in this way it (dis)solves the "problem of access" to mathematical entities.[41]

3 Platonism and constitution in Gödel's view

In the 50's Gödel writes that mathematical entities exist in a «second reality completely separated from space-time reality»,[42] that «concepts form an objective reality of their own»[43] and that «when I say [...] classes as objectively existing entities, I do indeed mean by that existence in the sense of ontological metaphysics».[44]

In the 60's, after the discovery of phenomenology, we find some changes in his writings, and some scholars[45] interpret this as a sign of the influence of Husserl's ideas.

In the revised edition of the article about Cantor's continuum problem Gödel writes that

> the question of the objective existence of the objects of mathematical intuition [...] is not decisive for the problem under discussion here. The mere psychological fact of the existence of an intuition which is sufficiently clear to produce the axioms of set theory and an open series of extensions of them suffices to give meaning to the question of the truth or falsity of propositions like Cantor's continuum hypothesis.[46]

Here Gödel introduces a new form of realism, called objectivism, centered not in the existence of mathematical entities, but in the idea that every proposition is either true or false, i.e. that every proposition (as Cantor's continuum hypothesis) has a determinate truth value. Wang calls

---

[41] Regard to the constitution and structure of mathematical ontology another subject should be treated, i.e. The relationship between construction and axiomatization, and the connected concept of incompleteness, as conceived by Gödel and by phenomenologists. This theme, however, would bring us out of the scope of this article.

[42] Gödel (*1953/9-III, 353).

[43] Gödel (*1951, 320).

[44] Gödel, letter to Gotthard Günther, june 30, 1954 (CW IV, 503-05).

[45] Føllesdal (1995), (1995a) and Hauser (2006).

[46] Gödel (1964, 268).

this conception of Gödel «objectivity over objects»,[47] and Kai Hauser interprets the shift from realism to objectivism as a sign of the transcendental reduction, but we have seen that the epochè does not put aside only existence but also truths. As Michael Dummett says,[48] objectivism is another form of realism (the only meaningful, in his opinion), and a constitutive approach, like the phenomenological one, cannot allow for the idea of a completely determined ontology, because the process of constitution is inexhaustible and therefore always "in progress": mathematical objects emerge from the vagueness of the life-world, and this vagueness is impossible to eliminate, so that we can in principle have some propositions without a determinate truth value, because the state of affairs they refer to is not (yet?[49]) determined. In this perspective incompleteness does not come only from the partiality of our intuition, as Gödel believes, but also from that of our constitution.[50]

Gödel seems, however, to give some importance to constitution in the following passage from Cantor's paper cited above:

> mathematical intuition need not be conceived of as a faculty giving an *immediate* knowledge of the objects concerned. Rather it seems that, as in the case of physical experience, we *form* our ideas also of those objects on the basis of something else which *is* immediately given. Only this something else here is *not*, or not primarily, the sensations. [...] It by no means follows, however, that the data of this second kind, because they cannot be associated with actions of certain things upon our sense organs, are something purely subjective, as Kant asserted. Rather they, too, may represent an aspect of objective reality, but, as opposed to sensations, their presence in us may be due to another kind of relationship between ourselves and reality.[51]

Is Gödel here welcoming here the thesis of a constitution of mathematical objects, perhaps because of the influence of Husserl, as

---

[47] Wang (1996, 242-46 and 303-305).

[48] Dummett (1978, Introduction). Gödel's objectivity over objects comes likely from Georg Kreisel, who was in close contact with Gödel in the late 50's. Kreisel wrote that Wittgenstein criticizes the idea of mathematical objects, but not that of mathematical objectivity (See, G. Kreisel, «Review of: Ludwig Wittggenstein's Remarks on the Foundations of Mathematics», British Journal for the Philosophy of Science, Vol. N. 9, 1958, p. 138 note 1). This is what in analytic philosophy is widely known as Kreisel's dictum, see for example Dummett (1978).

[49] In certain situations we do not know if it will be possible to eliminate vagueness from a state of affairs, and an example is the continuum problem. In other situations that seems to be even impossible, and an example is the indetermination of position and linear momentum in quantum mechanics.

[50] Bailly - Longo (2008, 268-69).

[51] Gödel (1964, 268).

Dagfinn Føllesdal suggests? He refers to phenomenological constitution in some conversations reported by Wang[52] but his words about it are generic and confused, and, what is more important, constitution, like formation in the passage cited above, acts «on the basis of something else which *is* immediately given» and which «is *not* [...] the sensations», but nevertheless «represents an aspect of objective reality».

Already in the 50's Gödel spoke about a process of construction of mathematical objects, but only as combining «some given material [...] then this material or basis for our constructions would be something objective and would force some realistic viewpoint upon us».[53] In the 70's, Gödel confirms this view to Wang: «creation in this sense does not exclude Platonism. It is not important which mathematical objects exist but that some of them do exist. Objects and concepts, or at least something in them, exist objectively and independently of the acts of human mind».[54]

Summarizing these passages, we notice a continuity from the Gibb's Lecture to the revised Cantor paper to the conversations with Wang: Gödel admits a process of construction – formation – constitution, but starting from some basic data which are objective (i.e. refer to reality) but are not sensations. These data of a «second kind» are basic ideas or concepts, like the «idea of object»,[55] and are given to us through «another kind of relationship between ourselves and reality» as opposed to sensation. Neither intuition nor this mysterious relationship can be Husserlian intentionality: we have seen that intentionality constitutes objects from the hyle, and the hyle is a complex of sense data (but Gödel's immediate data of the second kind are not sensations) reduced by the epochè, and therefore in no sense objective or existing (as Gödel's are).

These data of a second kind, according to Gödel, come from the Platonistic «second reality» of mathematics and are received by an «additional sense»,[56] which Gödel referred to as «reason»,[57] which recalls the Kantian noumenical faculty. As the data of the first kind, i.e. sensations, are the result of «actions of certain things upon our sense organs», the data of the second kind, i.e. basic ideas or concepts, are the result of an action of the ideal concepts of mathematics upon our reason. Starting with these two immediate data, our mind re-constructs objective reality in its two levels:

---

[52] Wang (1996, 256 – 8.2.8; 301 – 9.2.27).
[53] Gödel (*1951, 312).
[54] Wang (1996, 225 – 7.2.10).
[55] Gödel (1964, 268).
[56] Gödel (*1953/9-III, 354).
[57] Gödel (*1953/9-III, 354).

physical reality and mathematical reality. Thus, in Gödel's view, «creation in this sense does not exclude Platonism». [58].

Mathematical reality is not sensible and not material, but also in physics matter is illusory, and physical objects are composed by spiritual monads.[59] Thus Gödel denies materialism in two ways: the physical world is - beyond the appearences - spiritual, and the mathematical world has a formal and conceptual existence. Mathematical reality is also timeless, but time is illusory in physics as well, as Gödel argues in his paper about time and relativity.[60] As we have seen, however, according to phenomenology mathematical objects are "without matter" and "without time" in an entirely different sense.

4 Conclusion

The influence of Husserlian conceptions in the passages cited above (and in the whole of Gödel's philosophy) has been widely overestimated by thinkers like Føllesdal and Hauser. I consider better interpretations of Gödel's thought those given by Jairo da Silva and Giuseppe Longo. Da Silva stresses that Gödel «didn't pay enough attention to the fact that the inaugural act of the phenomenological attitude, the phenomenological reduction (*epochè*), was incompatible with the metaphysical theses he cherished»[61] and that he «never discussed the role played by transcendentally reduced consciousness in the constitution of [mathematical] objects».[62] Longo, then, contrasts Gödel and Husserl opposing «transcendence vs. transcendental constitution»:[63] «the difference», Longo remarks, «is given by the understanding of the object as constituted; it is not the existence of physical objects or of mathematical concepts that is at stake, but their constitution, as *their objectivity is entirely in their constitutive path*».[64]

Gödel considers mathematics a science of a real timeless world to which we have access via a special faculty known as "additional sense" or "reason", that makes possible a strong kind of intuition which reveals, at least partially, the structure of the "things in themselves". From a

---

[58] Wang (1996, 225 – 7.2.10)
[59] Wang (1996, 292-93 – 9.1.8-10).
[60] Gödel (1949a).
[61] Da Silva (2005, 554).
[62] Da Silva (2005, 566).
[63] Longo (2007, 207).
[64] Longo (2007, 209). According to Gödel, instead, objectivity rests in the basic data that we would receive from the "outer world".

phenomenological point of view, Gödel's metaphysical conceptions are totally unacceptable. As Longo says, «it is thus necessary to take Gödel's philosophy [...] and to turn it's head over heels, to bring it back to earth: one must not start 'from above', from objects, as being already constituted (existing), but from the constitutive process of these objects and concepts».[65]

References

I. Gödel's Works

The abbreviation CW, followed by the number of the volume, indicates Kurt Gödel's *Collected Works*:

CW I-V Gödel, K. (1986-2003): *Collected Works I-V*. S. Feferman et al. eds. Oxford, Oxford University Press.

The following abbreviations indicate the singles papers by Gödel, and are those already used in the *Collected Works.*

(1944): "Russell's Mathematical Logic". CW II, 119-141.

(*1946/9-B2): "Some observations about the relationship between theory of relativity and Kantian philosophy". CW III, 230-246.

(1947): "What is Cantor Continuum Problem?". CW II, 176-187.

(1949a): "A remark about the relationship between relativity theory and idealistic philosophy". CW II, 557-562.

(*1951): "Some basic theorems on the foundations of mathematics and their philosophical implications". CW III, 304-323.

(*1953/59-III): "Is mathematics syntax of language?". CW III, 334-356.

(*1961/?): "The modern development of the foundations of mathematics in the light of philosophy". CW III, 374-387.

(1964): "What is Cantor's continuum problem?". Revised and expanded version of 1947, CW II, 254-270.

II. Husserl's Works

The following abbreviations indicate Edmund Husserl's works. The abbreviation Hua, followed from the number of volume, indicates the

---

[65] Longo (2007, 209-10).

Husserliana critical edition published from 1950, by Nijhoff (Den Haag), and from 1989 on by Kluwer (Dordrecht – Boston – London).

(LU I-II): *Logische Untersuchungen I-II.* 1st Ed. 1900-01, Hua XVIII-XIX, 1975.

(IP): *Die Idee der Phänomenologie.* 1st Ed. 1907, Hua II, 1973.

(IDEEN): *Ideen zu einer reinen Phänomenologie und phänomenologischen Philosophie Erstes Buch.* 1st Ed. 1913, Hua III/1 e III/2, 1976.

(IDEEN II): *Ideen zu einer reinen Phänomenologie und phänomenologischen Philosophie. Zweites Buch.* Hua IV, 1952.

(CM): *Cartesianische Meditationen und Pariser Vorträge.* 1st Ed. 1931, Hua I, 1950.

(FTL): *Formale und transzendentale Logik.* 1st Ed. 1929, Hua XVII, 1974.

(KRISIS):*Die Krisis der europäischen Wissenschaften und die transzendentale Phänomenologie.* 1st Ed. 1936, Hua VI, 1954.

(EU): *Erfahrung und Urteil.* 1st Ed. 1938, Felix Meiner Verlag, Hamburg, 1999.

III. Secondary Literature

Bailly, F.; Longo, G. (2008): "Phenomenology of Incompleteness", in R. Lupacchini (Ed.), *Deduction, Computation, Experiment.* Milano, Springer, pp. 243-272. (http://www.di.ens.fr/users/longo/download.html)

Da Silva, J. (2005): "Gödel and Transcendental Phenomenology", in J. Hintikka (2005), pp. 553-580.

Dummett, M. (1978): *Truth and Other Enigmas.* Cambridge MA, Harvard University Press.

Føllesdal, D. (1995): "Introductory Note to *1961/?*". CW III pp. 364-373.

Føllesdal, D. (1995a): "Gödel and Husserl", in: J. Hintikka (Ed.), *From Dedekind to Gödel. Essays on the Development of the Foundations of Mathematics.* Dordrecht Springer, 1995, pp. 427-446; editet also in J. Petitot & C. (Eds.), 1999: *Naturalizing Phenomenology.* Stanford, Stanford University Press, 1999, pp. 385-401.

Gehlen A. (1940): *Der Mensch. Seine Natur und seine Stellung in der Welt.* Wiesbaden, Akademische Verlagsgesellschaft Athenaion.

Giusti, E. (1999): *Ipotesi sulla natura degli enti matematici*. Torino, Bollati Boringhieri.

Hauser, K. (2006): "Gödel's Program Revisited. Part I: The Turn to Phenomenology". *The Bulletin of Symbolic Logic*, Vol.12, N. 4, pp. 529-590.

Hintikka, J. (2005): Kurt Gödel. *Revue Internationale de Philosophie*, 4-2005, Vol. 59, n° 234.

Longo, G. (2007): "Mathematical Concepts and Physical Objects", in L. Boi, P. Kerzberg, F. Patras (eds.). *Rediscovering Phenomenology*. Dordrecht, Springer, pp. 195-228.
(http://www.di.ens.fr/users/longo/download.html)

Rabanaque L.R. (2005) "Why the Noema". *Phenomenology 2005*. Vol. 2 - *Selected Essays from Latin America* Part 2, (Loparic Z., Walto R. Eds.). Bucharest, Zeta Books, pp. 447-467.

Tragesser R. (1991): "How Mathematical Foundation all Come About: a Report on Studies Toward a Phenomenological Critique of Gödel's Views on Mathematical Intuition", in D. Føllesdal, J. N. Mohanty, T. M. Seebohm (Eds.), *Phenomenology and the Formal Sciences*. Dordrecht, Kluwer, pp. 195-213.

Wang, H. (1996): *A Logical Journey. From Gödel to Philosophy*. Cambridge, MIT Press.

How to Understand a Diagram
Looking Into (one of) the Practice(s) of Mathematics

Valeria Giardino
Institut Jean Nicod (CNRS-EHESS-ENS), Paris
Valeria.Giardino@ens.fr

1 Introduction

In recent years, an interest has grown among philosophers in the practice of science in general and of a very complex science such as mathematics in particular. The aim of this relatively new approach to mathematics is to uncover the mechanisms that lead to a mathematical result. To reach this goal, it is necessary to take into account the activities in which the working mathematiciansare engaged in almost every day. Mathematics is a very rich and complex enterprise, and is both a human activity and an historical phenomenon. For this reason, the objective of providing a general framework that wouldfit all kinds of mathematical practices has proved very difficult. A good strategy to tackle this issue is either to evaluate the cognitive tools that mathematicians have at their disposal in their research or to look for interesting case studies in the history of mathematics; in some cases, these two lines of research might overlap. What would therefore be an appropriate framework to describe a mathematical practice?

In a famous article, Quine invited philosophers to *naturalize* epistemology, but his invitation sounded suspicious to philosophers of mathematics, due to the empiricism that it presupposed.[1] Years later,

---

[1]Quine (1968).

Kitcher proposed an interesting naturalized framework for mathematics, according to which mathematical practice is defined as the quintuple <L, S, R, Q, M>.[2] In this quintuple, L represents the language used in the practice, S the set of accepted statements, R a collection of the forms of reasoning implied by being familiar with the practice, Q the open problems still to be solved and finally M the *metamathematical* views. Despite its being more specific to mathematics, Kitcher's naturalized framework is still of a limited application, since it describes mathematical practice in a very abstract and disembodied fashion: where are the working mathematicians? For this reason, Ferreiros recently imagined a new model, according to which Kitcher's quintuple becomes the *Framework* and gets integrated by a new element: the *Agent*.[3] To clarify, in Ferreiros' version of the quintuple, L does not necessary mean a formal language, because his Framework is intended to be directly applicable to the analysis of historical given practices. The Agent instead can be analyzed at different levels. First, it is a subject who has developed typical cognitive activities and basic practices. Furthermore, it can be considered as the *historical* actor, who has specific metamathematical views and research agendas. Finally, it can also stand for the *collective* subject in relation to some particular community.

Thinking about mathematics in terms of frameworks and agents opens the ground for what could be defined a *post-foundationalist* philosophy of mathematics, where the research is focused on questions such as 'what is a *mathematical practice*?', 'what is the relationship between mathematical practice and mathematical knowledge?', 'what is the relationship between mathematical practice and mathematical understanding?', and so on and so forth. These lines of research define a new branch of philosophy of mathematics, the philosophy of mathematical practice, which is still youngbut has already produced very interesting results.[4]

Talking about practices, one of the most common mathematical practices, which is found in the most ancient as well as in its most recent history of mathematics, is mathematicians' recourse to drawings, sketches, diagrams, figures, both in order to find the result of a mathematical problem and in order to explain to a novice the steps that have brought to this result. It is rather uncontroversial to ascribe to diagrams and figures the capacity of representing a good heuristics in problem solving; nevertheless, beside this superficial agreement, the effectiveness of these cognitive tools has not been considered so far of much philosophical interest. Heuristics is a matter of psychology, some might say. Nevertheless, let us assume that diagrams and

---

[2]Kitcher (1984).
[3]Ferreiros (2010).
[4] Cf. Mancosu (ed.) (2008).

figures are heuristically good. What does that mean? If a diagram or a figure is indeed such a good cognitive tool, are there rules that must be followed in order to use it? Do mathematicians have to learn how to use a figure or a diagram as a mathematical tool?

My attempt in this article will be to discuss the role that diagrams and figures play in mathematics in relation to the background knowledge – the Framework - and the mental abilities of the working mathematician – the Agent. In order to do that, I will present two examples that show how diagrams and figures are effective in leading to some mathematical result; secondly, I will consider the constraints diagrams are subject to and the activities in which mathematicians are engaged in when they make use of them; finally, I will present my view on how diagrammatic reasoning works and on the importance of learning a manipulation practice.

2 Two visual proofs

2.1. VP1: the Pythagorean Problem

As a first example of an effective mathematical diagram, let us consider one of the many visual proofs of a very well known mathematical theorem, the Pythagorean Theorem (PT). According to PT, as it is introduced by Euclid in the *Elements* (Book I, Proposition 47):

> (PT) *In right-angled triangles the square on the side opposite to the right angle equals the sum of the squares on the sides containing the right angle.*

There exists almost a hundred different proofs of the Pythagorean Theorem that rely on its visualization and on different ways of manipulating, moving, superimposing, coloring, slicing, rotating some of its elements.[5] Most of the times, people who are guided through these proofs seem to haveno difficulties in forming the belief that, thanks to the visual proof, PT holds. In some of the most extreme cases, realizing that the proof, unconventional as it may be, still gives them good reasons for believing that PT holds, will even surprise them.

In the following paragraph, I will guide the reader through one of these proofs - or better, through two slightly different versions of the same proof - that to my knowledge takes inspiration from some sketches and text

---

[5] Cf. Maor (2007).

by Leonard de Vinci. I will start from the instructions Euclid gives in the proof of PT that he offers in the *Elements*: given a right-angled triangle *ABC*, draw the squares on its sides, labeled *ADEB* and *BFGC*, and the square on the hypotenuse, labeled *IACJ*.

The two versions of the visual proof of PT that I will present make use of the same figure (Fig. 1) in a slightly different way. The first proof (*VP1*) will be done in five steps, and these steps are meant to construct the final and effective figure which is represented in Fig. 1; the second proof (*VP1'*) is instead done in just three steps, and provides a more dynamic and compressed description ofFig. 1, as I will show. As we know, the conclusion to obtain by means of the visual proof is that the area of the square built on the hypothenuse is the sum of the areas of the squares built on the small sides: $AC^2 = AB^2 + BC^2$.

Let us start with *VP1*:

1　Join the vertex *E* and *F* of squares *ADEB* and *BFGC* so that a second right triangle *EBF* is obtained: this triangle is congruent to the original one *ABC*; draw a third triangle which is a copy of *ABC*, but rotated 180°, such that its hypotenuse coincides with the inferior side of the square *ACJI* drawn on the hypotenuse of the triangle *ABC*.

2　Draw the two lines *BH* and *DG* that join, respectively, the opposite of the hexagons *DEFGCA* and *BCJHIA*.

3　By addiction of figures with the same areas, the four sided figure *DEFG* is congruent to the four sided figure *DACG*, and *BAIH* is congruent to *BCJH*; if *DACG* is rotated of 90° with A as a centre, it is shown to be congruent to *BAIH*.

4　Step 3 implies that the two hexagons *DEFGCA* and *BCJHIA* have congruent halves and, as a consequence, they also have the same area.

5　If twice the area of the triangle *ABC* is subtracted to each of these two hexagons, then *ACJI* =*ADEB* + *BFGC* and the theorem follows.

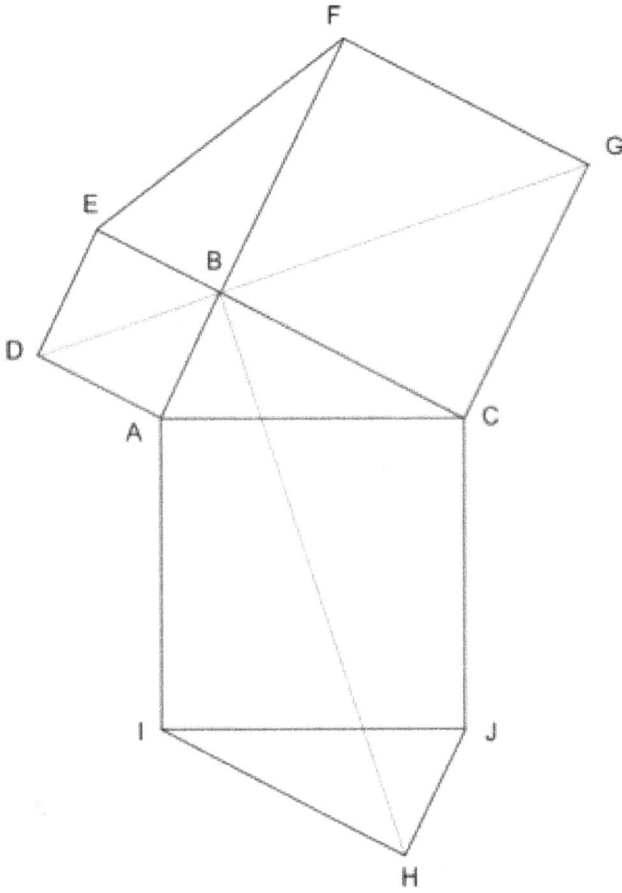

Figure 1

What *VP1* demands in order to obtain the conclusion is the construction of some elements into the original figure, the rotation of the figure, the superimposition of some of its elements, and finally the operation of adding and subtracting some parts of the figure to others.

Let us now have a look at *VP1'*, which replaces steps 3-5 with a single step 3':

3'. Rotate the figure 90° with *A* as centre, thus transforming *B* in *D*; send *AI* on *AC*, *AIH* on *ACG* and *IH* on *CG* and finally *BAIH* on *DACG*. This move implies that the area of *BAIH* is equal to the area of *DACG*, and the theorem follows.

The difference between *VP1* and *VP1'* is that *VP1'* emphasizes more the operation of rotation and superimposition than *VP1*, on which both proofs rely on. In fact, *VP1'* does not mention explicitly certain steps: in *VP1'*, for instance, it is not necessary to introduce the two hexagons *DEFGCA* and *BCJHIA* and to check for their congruence, since thanks to the rotation and the superposition, the congruence of *BAIH* and *DACG* is directly inferred.

2.2. VP2: the formula for the sum of the first natural numbers n

As a second example, let us consider the sum of the first natural numbers *n* (SN). According to the formula:

(SN) *The sum of the first natural numbers* n *is equal to* n *multiplied by* n *plus* 1 *and then divided by* 2.

$$S_n = \frac{n\,(n+1)}{2}$$

I will present a diagram that allows checking for this equivalence.

With respect to the formula above, the choice of the diagram to prove that it holds is crucial: the configuration chosen to arrange the data has an influence on the process of finding the solution. According to a well known story, little Gauss was given by his teacher the problem of finding the value of the sum of the numbers from 1 to 100, and he successfully and very quickly provided the answer by just realizing that the sum of the numbers equidistant from the middle was constant. The sum of the first 100 numbers is in fact easily solved as the sum of (1+100) + (2+99) + (3+98) and so on, that is 101 x 50. If we generalize this clever move, we get to the formula above. Also in the visual proof I will present, the diagram represents a good arrangement of the first natural numbers, in such a way that the structure of the answer is outlined.

As in the PT example, the visual proof of SN (*VP2*) as well is due in several steps. Each step is crucial in order to show that the left side of the formula,S_n, which is simply a shortcut for $1 + 2 + 3 + \ldots n$, is equivalent to its right side.

Let us follow the instructions that bring to figure 2 and consequently to *VP2*:

1. Arrange some circles in a rectangle so that the rectangle contains n $(n+1)$ circles.
2. Color half of the circles in black in the first half of the rectangle along the diagonal; the black circles will thus be n $(n+1)/2$.
3. Move the gaze from the upper angle on the left to the lower part of the rectangle in Fig. 2, and *see* now the black circles *as* $1+2+3+4+5+...+n$.
4. Therefore, $1+2+3+4+5+...+n = n(n+1)/2$.

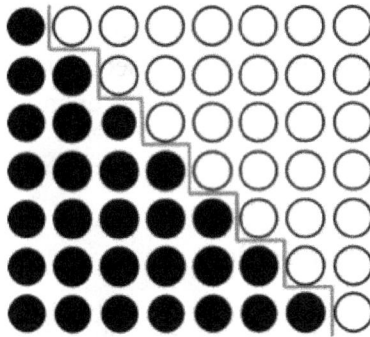

Figure 2

It does not really matter here whether the circles are of a particular number n as in the diagram in Fig. 2: by contrast, what is crucial here is their spatial arrangement. Whenever we go from a number n to a number $n+1$, the configuration will not be altered. What *VP2* demands in order to obtain the conclusion is the rule to obtain the area of a rectangle, the ability to count and, more interestingly, the capacity of *seeing* the black circles *as* half of the rectangle of sides n and $n+1$, and *at the same timeas* the very first n numbers.

3 Constraints and cognitive activities implied by VPs

By having been guided through *VP1*, *VP1'*, and *VP2*, the naive reader has maybe concluded that diagrams are easy to be used in order to find the good strategy to prove some mathematical result and to explain the reasons why such a result is obtained. This belief is most of the time expressed by locutions such as 'reading off' the information from a diagram, or extracting the information 'at zero costs'.

Nevertheless, it is necessary to carefully examine what makes itpossible for a diagram or a figure to convey information at all, and what are the characteristics of our reasoning through it. In fact, the capacity of understanding *VPs* does not come 'for free', as some of the enthusiastic literature about visual thinking seems to suggest.[6] Of course, diagrams are out there for us; nevertheless, some constraints apply to them. These constraintsare due to the fact that in reasoning with a diagram the user is not merely looking at an object, but more precisely she is engaged in a complex form of seeing that object, a diagram, *to the aim* of performing some inference and solving a mathematical task.

In the following sections, I will analyze diagrammatic reasoning from two points of view: first, I will move in the perspective of the *object* - what are the diagram's constraints? - and secondly in the perspective of the *subject* - in which activities is the reasoner engaged in when she is reasoning by means of the diagram?

3.1. The constraints on the object

Once again, a common intuition is that diagrams and figures convey information 'for free': the idea is that it suffices for the user to look at them to grasp the message they convey. Nevertheless, there are at least four constraints diagrams must be subject to in order to convey information in an interesting way. I will present these four constraints in turn.

(C_1) Diagrams must be *generic*:

Of course, figures are drawn on paper or shown on a computer screen. In general terms, the word *figure* is commonly used and even stands as a caption in texts such as this article, when figures are presented by counting them (Fig. 1, Fig. 2, and so on). Nevertheless, this term is in the end ambiguous, since it can be intended to refer both to a geometrical - that means *generic* – figure, and at the same time to the actual figure or diagram drawn on a piece of paper or shown on the computer screen that is its material two-dimensional representation. It is only the figure in the first sense that serves as a support for our reasoning about the mathematical problem. Such a figure is generic because it refers to a spatial arrangement that is created independently of the single drawing that shows it: in our examples, the generic figure is a 'generic' irregular and convex nine-sided polygon (*VP1* and *VP1'*) and a 'generic' structure in arranging some colored

---

[6]Larkin and Simon (1995).

circles *(VP2)*. If diagrams are not generic enough in relation to their aim, they will not be effective as tools.

(C$_2$) Some of the diagrams' properties must be *invariant*:

Once again, diagrams are drawn on a blackboard or traced by a stroke on the sand. Nevertheless, they must preserve some topological and geometrical properties in the space where they are placed. We can draw billions of right-angled triangles, as far as one of their properties, their having one angle which is 90° degrees, is invariant. The properties to be preserved in a diagram can vary and be more or less numerous. Manders proposes to distinguish in a diagram its *co-exact* conditions, which are «insensitive to the effects of a range of variation in diagram entries»and its *exact* conditions, which in contrast «would fail immediately upon almost any diagram variation».[7] This distinction is pertinent here. Also Euler, by introducing his famous circles as a useful representation to study syllogisms, was aware of the fact that these diagrams can unfold «all the mysteries of logic» and render the whole «sensible to the eye».[8] Nevertheless, this capacity does not simply depend on their appearance as circles: «we may employ, then, spaces formed at pleasure to represent every general notion, and mark the subject of a proposition by a space containing A, and the attribute by another which contains B». It is these 'spaces' that matter, not the particular figure they are of.

(C$_3$) Diagrams must be intended to a *particular aim*:

A crucial point that has been most of the time overlooked is that there are intentions behind diagrams: diagrams are given to solve a particular problem or to obtain a particular objective. For this reason, the user is requested not to lose the cognitive control to make the diagram work in relation to a *problem* of interest (of a geometrical interest in *VP1* and of a number theoretical interest in *VP2*). It is only once the problem has been defined that our visual exploration of the diagram omits accidental information and to discard non salient conditions - for example, by avoidingcounting the regions that are visible in the nine-sided polygon shown in Fig. 1, or by focusing on the circular shape of the items in Fig. 2.

(C$_4$) Diagrams must be *constructed* in steps:

The last constraint is on the process of construction of diagrams. In our examples, instructions were given to make the temporal order of the different construction steps explicit. If these instructions are appropriate, then the message in the diagram is about how to organize its space. Of course, the space can be *continuous* as in the PT example, or *discrete* as in

---

[7]Manders (2009)

[8]Letter 103, *Of Syllogism, and their different Forms, when the first Proposition is Universal*. See Euler (1997).

the SN one. The message a visual proof sends us is always about space, and therefore its nature is to some extent intrinsically geometric.

To sum up, there are four constraints: (C_1) *genericity*, (C_2) *invariance*, (C_3) *particular aim*, (C_4) *construction in time*.

3.2. The activity of the subject

The constraints on the object presented in the previous section match the capacities that the subject is requested to have in order to use diagrams to express a mathematical statement and together with it also its proof. What are the conditions of reasoning with diagrams on the side of the subject?

Here I will list the four activities that the user is engaged in when she appropriately uses a diagram:

(A_1) The user is *notsimplyseeing:*

In our examples, imagine moves like counting the regions that are recognizable in Fig. 1. That would mean pointing out at some properties of the figurethat are clearly *visible*; nevertheless, it will certainly not be considered as a good start in finding the proof of PT. There are things that the user *already knows*, such as the mathematical fact that 'a square must have 4 sides' and 'a right triangle must have a right angle', and these properties are immediately identified as such by perception. One interesting view on this issue isGiaquinto's analysis.[9] According to Giaquinto, the truth of the beliefs that the user forms in looking at a diagram is based on the fact that she already possesses resources that are *sufficient* to produce in her this belief, thanks to the visualization. Giaquinto's proposal is to consider several elements. In order to produce a belief which relies on the figure, the user must possess: (i) visual categories she can access; (ii) the correspondent perceptual concepts; (iii) the spontaneous capacity of finding connections in associative memory among these category specifications; (iv) verbal categories labels; (v) the geometrical concepts; (vi) a certain belief-forming disposition. It is only thanks to the possession of (i)-(vi) and of the concept of restricted universal quantification that this disposition is activated in the user and beliefs are formed. In our examples, the user has not empirical evidence for them, but knows them *a priori*; moreover, her reasoning process does not depend on meaning analysis and is not deduction from definitions, but is *synthetic*. Therefore, in Giaquinto's view, at least some geometrical knowledge isin the end*synthetic a priori*: even in the simplest use of a geometrical diagram, visual categorizing, verbal categorizing and belief-forming dispositions are all involved and intertwined.

---

[9]Giaquinto (2007).

(A_2) The user is not mechanically *reproducing:*

Let us go back to our examples. It is only when the user *understands*Fig. 1 and sees it as a figure of a right-angled triangle with the three squares drawn on its sides, that he can *reproduce* itin a non mechanical way and without 'damages'. The same happens for Fig. 2, which has to be seen as a diagram of the disposition of the first *n* numbers multiplied by 2. To quote Manders again,

> to show what happens when an exact condition fails, one must use a diagram in which it fails in an exaggerated way. In actual geometrical reasoning, providing demonstrative grounds for a given co-exact attribution might require *re-drawing the diagram.*[10]

(A_3) The user is seeing what it is *intended* to be seen:

The user, by looking at the diagrams, is having something like the geometrical experience - not simply perceptual - of seeing what she*has to.* To this aim, the user has to recognize that there is a sense in which the diagram has been *given as* a tool to be used in the proof, and therefore it respects C_3. The user must be familiar with the particular problem-solving context and with the strategy that is proposed by the diagram.

(A_4) The user is making the diagram *interact* with other formats:

Finally, it is necessary not to forget or underestimate the role of letters or labels in figures and diagrams, as the *A, B, C,* and so on in *VP1.* Their role is far from simply decorative but is on the contrary crucial: letters and labels of this kind elicit the user's attention. Interestingly, Peirce claimed that letters are *indexes*: an index is a sign which asserts «nothing; it only says 'There!' It takes hold of our eyes, as it were, and forcibly directs them to a particular object, and there it stops».[11] Of course, this makes sense only when the activity of the user is considered in her interaction with different formats and representations. The index constitutes a dynamical connection «both with the individual object, on the one hand, and with the senses or memory of the person for whom it serves as a sign, on the other hand».[12]

To sum up, there are four activities: (A_1) *no simple seeing,* (A_2) *no mechanical reproduction,* (A_3) *experience,* (A_4) *interaction among formats.*

---

[10]Manders (2008).

[11] Peirce (1885) .

[12] Peirce (1901).

4 Manipulation practices

Let us now go back to the issue about visual proofs: what is distinctive about them? The best way to answer to this question is to renounce the idea of a definition of what a diagram is in general, and look at the ways a subject*uses* a diagram, as I have just shown. We have arrived thus at the formulation of a set of constraints on the diagram and of a set of activities onthe user relying on *VPs*. This is a 'processing' approach: the user manipulates a diagram or applies to it some procedure that she has learnt as commonly accepted or that she believes that she is entitled to apply.

As a consequence, when we consider the distinctive nature of *VPs*, the point is to provide a framework where all the features that I have discussed so far could fit in.I will assume a pragmatic approach to figures and diagrams in mathematics. Grosholz claims that «an epistemology that works properly for mathematics will have to take into account the pragmatic as well as the syntactic and semantical features of representation in mathematics». Moreover, «different modes of representation in mathematics bring out different aspects of the items they aim to explain and precipitate with differing degrees of success and accuracy».[13] This pragmatic approach studies the use of the different scaffolding structures available in mathematics in terms of their representational role in an historical context of problem solving.

Imagine some particular diagrams that are offered as a tool to find the solution of some mathematical problem. There are of course all kinds of possible transformations these diagrams can be subject to. If they weresimply external scaffolding structures for their user's thoughts, then it would seem that the user is entitled to perform all sorts of actions in order to manipulate them. But is this actually the case? No. As I have explained in the previous section, not all transformations will indeed do. Why?

In a diagram, many different transformations are possible; nevertheless, when the user understands the diagram along the lines individuated by A_1 - A_4, the number of these possibilities is reduced. Moreover, because of C_1 - C_4, of all the set of possible transformations, only some of them are recognized as legitimate and as a good strategy to arrive at the solution of the problem.

To give you an idea of what I mean, let us go back to our examples and *see* how some manipulations are clearly considered legitimate, and others not.

---

[13]Grosholz (2007).

Let us begin with Fig. 1. In *VP1* and *VP1'*, the manipulation shown in Fig. 3 was performed: the figure was rotated of 90° degrees with A as the center of the rotation.

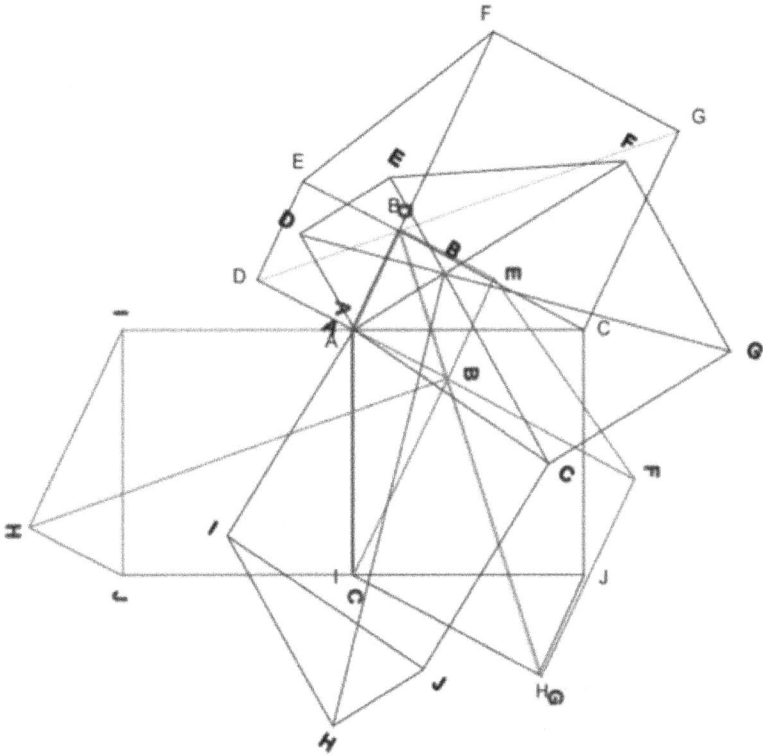

Figure 3

The user - and the reader with her I suppose - does not have any problem with performing and accepting such manipulation as appropriate. Nevertheless, the same user would not accept other kinds of manipulations, such as stretching the figure from its sides, as shown in Fig. 4.

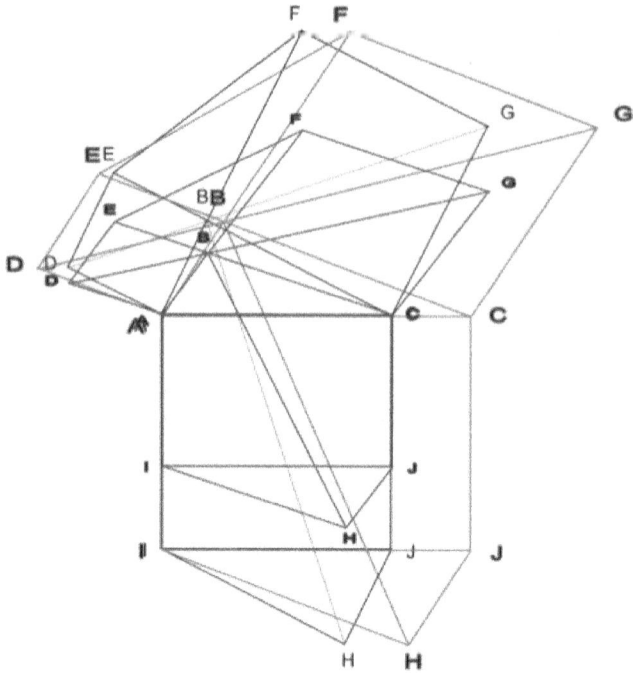

Figure 4

What is therefore the connection between these manipulations and the Cs or As implied? Let us consider for example C_1. The figure here is *generic*, because it refers to a spatial arrangement - a right-angled triangle and three squares on its sides - which is independent from the particular figure that is now printed on this page. Moreover, because of C_2, it is precisely these properties - the one of showing a right-angledtriangle and three squares, that need to be *invariant*, and therefore the manipulation in Fig. 4 is not available. For what regards the activities the user is engaged in, let us consider A_1. As I have already said, the user could surely count the regions that are recognizable in Fig. 3 and in Fig. 4, but this move will not be recognized as the appropriate way of looking at these figures. It is not because the regions in Fig. 3 and Fig. 4 are different that the manipulation in Fig. 4 does not apply. By contrast, it is necessary to recognize once again which visual properties of the figure must be invariant. For this reason, if the user is asked to reproduce the figure in figure 1, she will do it in a non mechanical way, and she will pay attention at drawing it without 'damaging' the original.

Therefore, the operation of stretching the right triangle is not available in the *VP1* case; by contrast, this manipulation applies in the *VP2* case. Stretching the diagram by its sides will not create any difficulty in preserving the configuration, as shown in Fig. 5: the spatial arrangement that the diagram is displaying and that constitutes the support for our reasoning will not be altered.

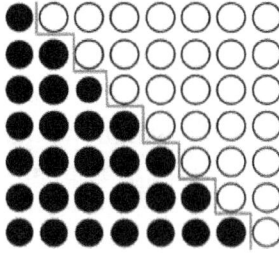

Figure 5

Other kinds of manipulations such as adding new elements to the diagram of Fig. 2 will do only in some cases. In both Fig. 6(a) and Fig. 6(b), new circles different from the original black and white ones are added. Nevertheless, the outcome is not the same. In Fig. 6(a)the spatial invariances are respected: even if the new circles are grey, some of them - the circles on the right - have a black boundary that is enough to distinguish them from the circles on the left and therefore preserve the sequence in the original diagram. By contrast, in Fig. 6(b), such a structure is not respected. The added circles alter the original diagram; the new diagram ceases to be a useful support for our reasoning on SN.

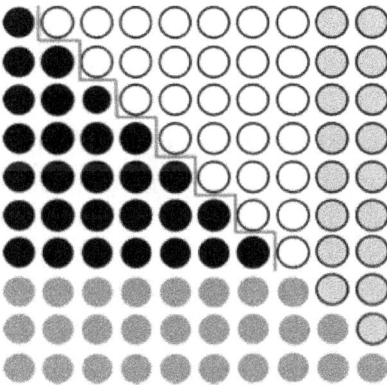

Figure 6(a)

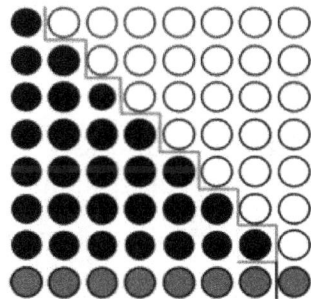

Figure 6(b)

If we consider C_3, the spatial arrangement in these diagrams is *intended* to display the first natural numbers. If this is the relevant information, then of course this aim will be still met in Fig. 5 and Fig. 6(a), but lost in Fig. 6(b). In fact, a mistake has been made in one of the successive *steps* in the instructions on how to draw the figure in Fig. 6(b). This information, as C_4 prescribes, could be extracted from Fig. 2. For what concerns the activity of the subject, we have already said in discussing C_3 that as A_3 indicates, the user has to see what it is *intended to be seen.*Furthermore, in this example it is particularly clear the way in which we have an *interaction* with other formats, as proposed in A_4. In fact, it is precisely the formula that is meant to be proved that suggests to the user to *see* the diagram *first*as the picture of what is indicated in the left side of the formula - the sum of the first natural numbers - and *afterwords*as the picture of what is indicated in the right side of the formula - half a rectangle of sides n and $n + 1$.

What I am pointing out at here in reconsidering VP1 and VP2 in such a way, is the importance of the knowledge and the familiarity of manipulation practices in accordance with the used system of representation. The user *sees*Fig. 3, Fig. 5 and Fig. 6(a)*as* legitimate manipulations and by contrast Fig. 4 and Fig. 6(b)*as* aberrations. Nevertheless, this happens not because she has learnt some explicit rule that settles which manipulations are accepted and which are not, but because she is familiar with a manipulation practice that applies to these particular sorts of diagrams. The familiarity with these practices is a crucial aspect of diagrammatic reasoning in general and is even more crucial in mathematics, where these practices determine our inference capacity and the extent to which diagrams can serve as external scaffolding structures for our reasoning.

5 Conclusions

In this article, I discussed the relations that connect diagrams, mathematical expertise and instructions to organize space. To conclude, consider the triangle in Fig. 7.

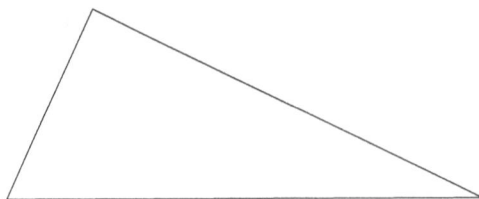

Figure 7

In the *Philosophical Investigations*, Wittgenstein claims that this triangle can be seen

> as a triangular hole, as a solid, as a geometrical drawing; as standing on its base, as hanging from its apex; as a mountain, as a wedge, as an arrow or pointer, as an overturned object which is meant to stand on the shorter side of the right angle, as a half parallelogram, and as various other things.[14]

In his remarks, Wittgenstein points out that language has the role here of disambiguating this figure, solving the polyvalence just shown: in describing - even only naming - the figure, it is not possible to attribute all these meanings to the figure *at the same time*. By naming or describing the figure, the user has to make a choice.

What I am suggesting here is that when we take into account the role of diagrams to support inferences about the diagram's message, these linguistic descriptions will give the instructions for organizing the space on the piece of paper or on the computer screen. These descriptions will in fact correspond to the application (or misapplication) of some manipulation practice, that is some legitimate (or illegitimate) manipulation of the diagram. It is only by means of these manipulations that the user shows the way she reacts to the diagram, and thus the way she reasons *with* it.

Let us suppose that a user makes a choice and decides to change the Wittgensteinian triangle as shown in Fig. 8.

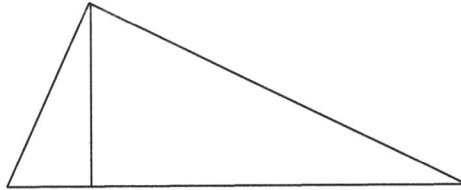

Figure 8

We are then almost back where we have started, that is to some generalization of PT, call it PT'.It's Euclid again, Book VI, Proposition 31:

> (PT') *In right-angled triangles the figure on the side opposite the right angle equals the sum of the similar and*

---

[14] Wittgenstein (1953/2001), Part II.

similarly described figures on the sides containing the right angle.

Since Euclid has already proved that similar triangles are to one another in the duplicate ratio of the corresponding sides (Book VI, Proposition 19), then we can substitute 'figure' with 'triangle'; Fig. 8 shows that PT' holds, only if the user *sees* (in the sense we have discussed) that the triangles on the sides can be not only outside the original triangle, but also inside it.

Acknowledgements

Much of what I claim in this article is the outcome of my joint work and discussions with Mario Piazza. I thank Pierluigi Graziani and Massimo Sangoi for the organization of the workshop on Open Problems in the Philosophy of Science. The research was supported by the European Community's Seventh Framework Program ([FP7/2007-2013] under a Marie Curie Intra-European Fellowship for Career Development, contract number N° 220686 - DBR(Diagram-based Reasoning).

References

Euler, L. (1997): *Letters of Euler to a German Princess: On Different Subjects in Physics and Philosophy*. Translated by H. Hunter. London and New York, Thoemmes Continuum.

Ferreiros, J. (2010): "Mathematical Knowledge and the Interplay of Practices", in M. Suárez, M. Dorato, M. Rédei (eds.), *EPSA Philosophical Issues in the Sciences, Launch of the European Philosophy of Science Association*. Berlin, Springer.

Giaquinto, M. (2007): *Visual Thinking in Mathematics*. Oxford, Oxford University Press.

Giardino, V.; Piazza, M. (2008): *Senza parole. Ragionare con le Immagini.* Milano, Bompiani.

Grosholz, E. (2007): *Representation and Productive Ambiguity in Mathematics and the Sciences*. Oxford, Oxford University Press.

Kitcher, P. (1984): *The Nature of Mathematical Knowledge*. Oxford, Oxford University Press.

Larkin, J.; Simon, H.A. (1995): "Why a Diagram is (Sometimes) Worth Ten Thousand Words", in B. Chandrasekaran, J. Glasgow and N. H. Narayan (eds.), *Diagrammatic Reasoning: Cognitive and Computational Perspectives*. Menlo Park, California, AAAI Press; The MIT Press, pp. 69-109.

Mancosu, P. (2008): *The Philosophy of Mathematical Practice*. Oxford, Oxford University Press.

Manders, K. (2008): "Diagram-Based Geometric Practice", in P. Mancosu (ed.), *The Philosophy of Mathematical Practice*. Oxford, Oxford University Press, pp. 65-79.

Maor, E. (2007): *The Pythagorean Theorem*. Princeton, Princenton University Press.

Peirce, C.S. (1885): "On the Algebra of Logic: A Contribution to the Philosophy of Notation", in *The American Journal of Mathematics*, 7 (2): 180–202; reprinted in *Collected Papers of Charles Sanders Peirce* 3.359-3.403.

Peirce, C.S. (1901): *Dictionary of Philosophy & Psychology vol. 1*, in *Collected Papers of Charles Sanders Peirce,* 2.305.

Peirce, C.S. (1931-1958): Collected Papers of Charles Sanders Peirce, 8 vols. Edited by C. Hartshorne, P. Weiss, and A. W. Burks. Cambridge Massachusetts, Harvard University Press. Vols. 1–6 edited by C. Harteshorne and P. Weiss, 1931-1935; vols. 7–8 edited by A. W. Burks, 1958.

Quine, W. van O. (1968): "Epistemology Naturalized" in *Ontological Relativity and Other Essays* (1969). New York and London, Columbia University Press, pp. 69-90.

Wittgenstein, L. (1953/2001), *Philosophical Investigations.* Oxford, Blackwell Publishing.

On the Very Idea of an Indispensability Argument[1]

Andrea Sereni
Vita-Salute San Raffaele University, Milan
sereni.andrea@unisr.it

1 *The* Indispensability Argument?

A vast part of the philosophy of mathematics of the past (and present) century has been occupied by the dispute between platonists (those who believe in the existence of – allegedly abstract – mathematical objects) and nominalists (those who deny the existence of mathematical objects). In the past few decades, an increasing amount of works has been devoted to a specific argument for platonism, the so-called Indispensability Argument (henceforth, IA).

The basic idea underlying IA is easy to state. IA stems from the rather uncontroversial acknowledgment that mathematical theories are commonly employed in the formulation of our best (i.e. well-confirmed) scientific theories.[2] Let us assume that we are justified in taking these scientific theories to be (at least approximately) true. If we further assume that we are

---

[1]   Much of what is presented in this paper originates in joint work with Marco Panza (IHPST, CNRS Paris 1), and can be found in more detailed form in Panza, Sereni (forthcoming). Earlier versions were presented on several occasions (in Paris, Bergamo, Bologna, Frankfurt, Padua, Parma and Nancy). I would like to thank the all audiences for helpful comments. Special thanks go to the organizers and to the audience of the *Open Problems in The Philosophy of Science* conference (Cesena, April 15-17, 2010).

[2]   'Scientific theory' is here understood as covering only empirical – and possibly social – sciences, but neither geometry nor mathematics.

justified in taking these scientific theories as true only if we are justified in taking as true the mathematical theories that we cannot avoid employing in formulating them, we can conclude that we are justified in taking those indispensable mathematical theories to be true and their mathematical objects to exist. Most of IA's success is due to its appealing to considerations that should acceptable to both parties in the debate, and to the fact that it seems to deliver a platonist conclusion on *a posteriori* grounds.

Appearances notwithstanding, it is far from obvious how this core idea should be spelled out in details if it is to give rise to a properly formulated valid argument. Since the basic structure of the argument has been suggested by Quine in many scattered remarks, and first explicitly appealed to by Putnam (1971), many refer to what is often labeled "the Quine-Putnam Indispensability Argument". However, formulations of IA differ vastly among commentators. It is therefore legitimate to seek for a minimal version of IA – minimal, that is, in so far as it features the fewest and/or less controversial premises needed in order to get the required conclusion – such that most of the versions on the market could be seen as different ways of improving upon it.

In order to attain this goal, we will have to assess whether the argument's conclusion can be reached even if some of the assumptions commonly assumed as essential to it are disregarded. A substantial consequence of this methodological inquiry is that some of these theses can, as a matter of fact, be dispensed with. And that more than one conclusion might be at stake.

2 IA: a sketch of the debate

Most discussions of IA begin by reporting Putnam's (1971) famous formulation:

Putnam's Indispensability Argument [PIA]
So far I have been developing an argument for realism roughly along the following lines: quantification over mathematical entities is indispensable for science, both formal and physical, therefore we should accept such quantification; but this commits us to accepting the existence of the mathematical entities in question. (p. 347)

Putnam's argument is based essentially on two notions: indispensability and quantification. Appeal to quantification clearly displays Putnam's adoption of Quine's criterion of ontological commitment [QC].[3] The argument is meant to establish what Field (1982, p. 50) would call «theoretical indispensability», i.e. the claim that quantification over certain mathematical objects is required in order to *state* scientific laws (cf. Putnam 1971, pp. 346-7).[4] Provided we accept that the relevant scientific theories are (at least approximately) true, and provided we accept a standard reading of [QC], [PIA] tells us that we have reasons for believing that (certain) mathematical objects exist. Given these provisos, it is an argument for platonism.

More recent discussion on IA, however, has centred on other notions beside indispensability and quantification. It is generally agreed that IA relies on other important theses of Quinean provenance, i.e. confirmational holism and naturalism. Both are controversial and multifarious theses, but we can state them in a short form for our convenience:

[CH] *Confirmational Holism*: empirical evidence does not confirm scientific hypotheses in isolation, but rather scientific theories as a whole. As a consequence, with respects to ontology, we are justified in acknowledging the existence of *all* those entities that are quantified over in our true or well-confirmed scientific theories.

[NAT] *Naturalism*: scientific theories are the only source of genuine knowledge. As a consequence, with respect to ontology, we are justified in acknowledging the existence *only* of those entities that are quantified over in our true or well-confirmed scientific theories.

At present, the most discussed version of IA that is faithful to the idea that the argument appeals also to [CH] and [NAT] is Colyvan's:[5]

---

[3] With important qualifications: see below, section 4. [QC], in short, is the claim that the ontological commitment of a theory is given by the objects that must be counted among the values of the variables of the existentially quantified statements that are entailed by the theory. *Locus classicus* is Quine (1948).

[4] Theoretical indispensability is different from indispensability for the derivation of theorems. Field (1980) needs accordingly two different strategies in order to neutralize both. Notice that Putnam (1956) contains a clear acknowledgement that mathematics is conservative in more or less Field's terms, and that this accounts for dispensability for derivations. Putnam, however, never believed in the theoretical dispensability of mathematics (cf. also Field 1980, note 18, pp. 112-13).

[5] Cf. Colyvan (2001 p. 11). Cf. also Resnik (2005, p. 430).

Colyvan's Indispensability Argument [CIA]

    i) We ought to have ontological commitment to all and only those entities that are indispensable to our best scientific theories;

    ii) Mathematical entities are indispensable to our best scientific theories;

[CIA] ------------------------------

    iii) We ought to have ontological commitment to mathematical entities

According to Colyvan, "the crucial first premise follows from the doctrines of *naturalism* and *holism*" (2001, p. 12; emphasis in the original), respectively as regards the 'only' and the 'all' directions. The above formulations of the two relevant theses seem to motivate this claim (which, as we shall see, is far from saying that the two theses must necessarily be assumed if the argument is to go through).

Most of the recent debate on IA has focused on the tenability of [CIA], in particular as regards its naturalistic and holistic alleged background assumptions. Maddy (1992, 2007) has pointed to clashes between the notions of holism and naturalism, concerning aspects of scientific practice (e.g. idealizations), that would make [CIA] unsound. Sober (1993) has argued against [CIA] that empirical evidence cannot even indirectly confirm mathematical theories. Both supporters and critics have generally taken naturalism and holism as essential to IA. Some attempted to offer weaker formulations. Holism has been the preferred target. Both Resnik (1995) and Dieveney (2007) have suggested ways of dispensing with holism, and even Colyvan suggests that «as a matter of fact, the argument can be made to stand without confirmational holism» (2001, p. 37).

More recently, much attention has been devoted to issues concerning the explanatory role that mathematical theories and mathematical entities are supposed to play in our understanding of empirical phenomena. Baker (2009, p. 613) has claimed that apart from indispensability «it needs to be shown that reference to mathematical objects sometimes plays an *explanatory role* in science», and has suggested an «enhanced» version of IA, intended to account for the role mathematical explanation can play in IA. The bearing of the notion of explanation in this context was already suggested in Field (1989, p. 14), and it squares nicely with the view – at which both Quine and Putnam hinted – that IA shares some important aspects with arguments for scientific realism based on inference to the best explanation: the existence of both theoretical entities and mathematical objects would be justified once it is acknowledged that they contribute in an essential way to the explanatory power of our scientific theories.

At this point one could wonder whether all this conceptual machinery is really needed in order to get the desired conclusion. And this seems to deserve a negative answer.

3 Indispensability without holism and naturalism

Let us consider the following argument:

i) We are justified in believing in believing some scientific theories to be true;
   [We are justified in believing T true]
ii) Among them, some are such that some mathematical theories are indispensable to them;
   [M is indispensable to T]
iii) We are justified in believing true these scientific theories only if we are justified in believing true the mathematical theories that are indispensable to them;
   [We are justified in believing T true only if we are justified in believing M true]
[MIA$_a$]-----------------------------
iv) We are justified in believing true that the mathematical theories indispensable to these scientific theories.
   [We are justified in believing M true]
v) We are justified in believing true a mathematical theory only if we are justified in believing that the objects in the domain of the objectual quantifiers of the mathematical theory exist.
   [We are justified in believing that M is true only if we are justified in believing that the objects quantified over in M exist]
[MIA$_b$]-----------------------------
vi) Thus, we are justified in believing that the objects in the domain of the objectual quantifiers of the indispensable mathematical theories exist.
   [We are justified in believing that the objects quantified over in M exist]

It seems uncontroversial to take [MIA] as a valid version of IA. Whether it is also a sound one is something that need not interest us here. It remains to be established in what it differs from other formulations on the market.[6]

[MIA] recalls the argument structure of Putnam's [PIA]. Premise *(ii)* appeals to the notion of indispensability, whereas premise *(iv)* follows from Quine's criterion of ontological commitment. On its surface, [MIA] makes no appeal either to [CH] or to [NAT].[7] To say the least, [MIA] does not

---

[6] The explanatory role of mathematics can accounted for once we realize that the notion of indispensability is relational in character. We can define the notion of (in)dispensability as follows: a theory M is dispensable from a given theory T if and only if there is a theory T' that does not include statements in which the vocabulary of M occurs and that: *a)* is ε-equivalent to T, where ε is an appropriate equivalence relation; *b)* is equally or more virtuous than T according to an appropriate criterion of virtuosity. If T includes statements in which the vocabulary of M occurs, and there is no theory T' satisfying the above conditions, then M is indispensable to T. Once the notion of (in)dispensability is thus defined, the explanatory role of mathematics can be accounted for by the selection of an appropriate equivalence relation, such as sameness of explanatory power. For details and discussion, cf. Panza, Sereni (forthcoming).

[7] [MIA] is stated in epistemic terms: as it happens in [CIA] and elsewhere, it speaks of our justification in believing certain theories to be true and certain objects to exist. It is easy

feature anything similar to the "all and only" clause of [CIA]'s first premise. If something on the lines of [CH] and [NAT] is thought to be *expressed* by the "all and only" clause, then it is clear that neither thesis is expressed by the premises of [MIA]. Nevertheless, it might still be the case that [CH] and [NAT] are required in order to *justify* some of [MIA]'s premises. In particular, [NAT] might be needed in order to justify premise (*i*), whereas [CH] might be needed in order to justify premise (*iii*). It seems, however, that even this can be denied.

Much more than what can be said here would be needed in order to show that [CH] is not necessary for justifying premise (*iii*). Here the following will suffice. It seems that in order to claim that [CH] is a necessary condition for justifying premise (*iii*) one would be bound to *identify* empirical confirmation and justification: a scientific theory would thus be justified only to the extent that it is empirically confirmed. But this identification is all but uncontroversial. For one thing, if justification is here understood, as it seems required, as justification in believing something true, then saying that confirmation equals justification flies in the face of many account of confirmation according to which empirical confirmation falls short of delivering full-fledged justification in a theory (if it delivers justification at all). Second and foremost, it is agreed by most that a whole host of theoretical virtues can be relevant to the justification of a scientific theory, e.g. simplicity, familiarity of principles, explanatory power, unificatory power, and the like. Identifying justification with confirmation simply amounts to underestimating any vexed question of the underdetermination of theory by evidence, and to disregarding the contribution of other theoretical virtues. In general, thus, it seems that the condition expressed in premise (*iii*) can be accepted, and maybe even should, with no need of espousing an holistic conception of confirmation.[8]

---

to formulate a non-epistemic version of [MIA] simply by eliminating every occurrence of the sentential operator "we are justified in believing that". Given the way we have stated [CH] and [NAT], i.e. as these concerning justification, [CH] and [NAT] will not be relevant for the non-epistemic version of [MIA].

[8] As regards Putnam's own views, a precise assessment is all but uncontroversial (for some interpretations, cf. Liggins, 2008 and Marcus, 2010). Nonetheless, though Putnam acknowledged (e.g. in Putnam, 1979/1994) important connections between his general views and Quine's holistic picture of confirmation, holism never really seemed to underlie his adoption of IA, as the following passage witnesses: «I have never claimed that mathematics is "confirmed" by its applications in physics (although I argued in "What is Mathematical Truth" [Putnam (1975)] that there is a sort of quasi-empirical confirmation of mathematical conjectures *within* mathematics itself)». Putnam's use of the term "quasi-empirical" has some, but not too tight, connections with the thought of Lakatos (and thus possibly of Polya, cf. Motterlini, 2002), as he himself seems to acknowledge (Putnam 1979/1994, p. 28).

That IA can be made to stand without holism is, however, something that has been already suggested, as we have already said. No-one seems on the contrary to believe that the same could be claimed of naturalism. Still, nothing seems to indicate that naturalism is required in order to justify premise (*i*). What is needed in order to justify that premise is rather some form of scientific realism, and this is a weaker position than [NAT]. Scientific realism sees scientific theories as *a* genuine source of knowledge, but need not consider them as *the only* genuine source. Scientific realism might be defined in a variety of ways. Putnam (1971, p. 338) assumes «that one of our important purposes in doing physics is to try to state 'true or very nearly true' (the phrase is Newton's) laws, and not only to build bridges or predict experiences». Following Psillos (1999, p. xix) we can define scientific realism as the conjunction of three stances: a metaphysical one, i.e. that «the world has a definite and mind-independent natural-kind structure»; a semantic one, that «takes scientific theories at face-value, seeing them as truth-conditioned descriptions of their intended domain, both observable and unobservable [...] capable of being true or false»; and an epistemic one, that «regards mature and predictively successful scientific theories as well-confirmed and approximately true of the world».[9] Nothing in this threefold description entails a naturalistic stance dictating that scientific theories are the only ones that can legitimately afford us genuine knowledge about the world (possibly, but not necessarily, understood as comprising mental, religious, emotional or social aspects).[10] Nothing in principle seems to forbid those who believe in the *a priori* character of mathematical knowledge the acceptance of [MIA] as a valid argument for platonism (although they will see it as a rather weak, and possibly superfluous, tool in contrast to other, *a priori* reasonings).

Clearly, one who independently holds either [CH] or [NAT] as part of one's philosophical framework will likely offer a version of IA having either theses among its (possibly implicit) assumptions. But this is not mandated. Moreover, If IA is made to appeal to both [CH] *and* [NAT], it provides sufficient and *necessary* conditions for its conclusion. It follows that its conclusion has no bearing on all those mathematical theories that do not find application in true or well-confirmed scientific theories. Quine

---

[9] Adding 'approximately' as a qualification in (*i*) and (*ii*) might call for some qualification in the exposition of the argument, but does not affect our present discussion in significant ways.

[10] No naturalist assumption is present in Putnam's [PIA]. As Putnam (forthcoming) stresses, the 'only' direction of premise (*i*) of [CIA] expresses a thesis he "never subscribed to in [his] life".

accepted this conclusion,[11] but a more plausible version of IA might want to avoid it. As a matter of fact, [MIA] does avoid it.

4 Indispensability without platonism

So far, we have granted that IA is an argument for platonism. Surely, this is the conclusion Quine pointed at, and the one that seems suggested by Putnam's [PIA]. It is also clearly the conclusion to which Colyvan aims with [CIA], to which most supporters of IA also aim, and that even most critics of IA take to be the obvious conclusion of the argument.

In recent times, however, some have started to doubt that IA can be effective as an argument in support of platonism. These doubts are commonly expressed by pointing out that IA (in one's preferred formulation) proves to suffer from important shortcomings.

A first doubt, raised by Baker (2003), can be expressed by recalling Benacerraf's (1965) argument. Suppose that we want to account for the indispensability of Peano Arithmetic for a given scientific theory by claiming that the set-theory to which Peano Arithmetic reduces is indispensable: we will be justified in taking set theory to be true and sets to exist. But which sets? If nothing in the application of set theory to our particular scientific theory mandates that one particular kind of sets (e.g. Von Neumann's) will have to be preferred to a rival kind (e.g. Zermelo's), IA is of no help in deciding between the two (or more) rival ontologies. Baker further notices that if one allows (even only as a working hypothesis) that there might be alternative foundations of the parts of mathematics that find application in science, alternative that is to set-theory – as category theory might be said to be – then indeterminacy is brought a step further: IA will also be powerless in discriminating among the alternative ontologies of set-theory and, e.g. category theory. Baker concludes by claiming that if IA turns out as a sound argument, it will surely be effective in rejecting nominalism, but the problem of the multiple realizability of mathematics «presents an insuperable obstacle [...][for] even if mathematics is indispensable to science, no particular collection of mathematical objects is indispensable».

---

[11] Cf. Quine (1986, p. 400) and Quine (1995, pp. 56-57). For discussions, cf. Parsons (1978), Maddy (1992), Leng (2002), Colyvan (2007). This conclusion concerning unapplied mathematics seems to fly in the face of a standard conception of platonism (if anything like this exist). Borrowing Putnam's term, we might say that IA is rather an argument for a "quasi-empirical" platonism (cf. fn. 9).

Pincock (2004) has pointed to a further and more general limitation of IA as concerns its platonist conclusion, focusing in particular on [CIA]. According to Pincock, Colyvan's (and others') idea of establishing a platonist conclusion on the basis (among other things) of the successful applications of mathematics hides the assumption of a specific account of how mathematics gets applied. According to this account, the existence of mathematical objects is what allows the application of mathematics in e.g. physics. But this is a questionable account. According to the alternative "mapping-account" that Pincock himself offers, «applications depend only on the relations between mathematical objects; but the realist-nominalist debate concerns the non-relational or 'intrinsic' properties of mathematical objects» (p. 70).[12]

More recently, Paseu (2007) has come to conclusions similar to Pincock's, but from a wholly different route. Paseu wonders whether scientific standards for the acceptability of theories can not just sustain our belief that mathematics is true, but also impose some particular metaphysics for it. In order to answer this twofold question, Paseu reviews a number of criteria commonly regarded by the scientific community to ground acceptability of a scientific theory (from publication in scientific journals up to simplicity and scientific fertility). Paseu answers affirmatively to the first half of his question: on the basis of the applicability and utility of mathematics in the sciences, scientific standards justify us in the belief that our mathematical theories are true. This would answer to what he calls "the pragmatic objection" (to any form of mathematical platonism based on considerations about the applications of mathematics), according to which «scientists, *qua* scientists, are not epistemically committed to the mathematics they deploy». But this leaves unanswered what he calls "the indifference objection", which «maintains that scientific standards endorse mathematics in the proper epistemic sense but that they do not endorse *platonist* mathematics» (p. 131, emphasis in the original). According to Paseu, the acceptance of mathematics in ordinary scientific practice does not lend support to any *particular* interpretation of the content of the relevant mathematical statements, be it a platonist or an anti-platonist one.[13]

What do these three views point at? In a way, the problem of the multiple reducibility of mathematics might constitute a case of its own. And it needs not be seen as a fatal objection to IA. Most platonist positions

---

[12] An evaluation of Pincock's account is beyond the scope of the present work. For subsequent debate, cf. Bueno and Colyvan (2011), Batterman (2010), Pincock (2011).

[13] Paseu (2007), pp. 148-9, gives some suggestion as to how, according to him, a mathematical statement could be held true even when no specific interpretation of its content is endorsed.

attempting to rescue platonism from the epistemological problems signalled by Benacerraf (1965, 1973) have tried to give an appropriate characterization of mathematical objects, to establish what conditions their existence should satisfy, and to show how an appropriate epistemology for these objects can be opposed to the nominalist's qualms. Part of the success of IA in the platonism/nominalism debate, however, is arguably due to the fact that IA can deliver its conclusion while remaining entirely neutral as to the nature of mathematical objects (leaving a characterization of the latter to independent arguments). Whether this metaphysical neutrality, as we might call it, of IA is to be seen as a virtue or a vice, is likely to depend on the overall philosophical framework in the context of which IA is being appealed to.

What about the views of Pincock and Paseu? Might these be seen as substantial objections to IA? It does not seem so. Let us distinguish *ontological realism*, i.e. the thesis that certain objects exist, from *semantic realism*, i.e. the thesis that certain statements are true, without specific commitment to what makes them true.[14] When mathematical discourse is at stake, ontological realism is platonism. Is there any conceptual need of conflating mathematical ontological and semantical realism? Obviously thee is none. On the contrary, many semantic realist views are neatly distinguished from platonist ones: think e.g. of Putnam's (1967) theory of equivalent descriptions, or Hellman's (1989) modal structuralism. All those who believe with Kreisel that the real problem of mathematics is not that of accounting for the existence of mathematical objects but rather for the objectivity of mathematical statements, might welcome a version of IA that delivers the latter as a consequence without delivering the former.

Let us now come back to [MIA], then. What we get from [MIA] if we eschew premise (*v*) and thus conclusion (*vi*) (i.e. if we stop at the sub-argument [MIA_a]) is exactly a version of IA whose conclusion is *just* semantic realism about mathematics, short of platonism. Notice that this is not simply meant to be a formal trick. It rather points to the fact that a valid minimal indispensability argument can be devised that respects many (semantic) realist though anti-platonist intuitions in the philosophy of mathematics, from which arguments for platonism can, but need not, be obtained by the addition of extra premises.

Notice, in passing, that once a distinction between [MIA_a] and [MIA_b] is acknowledged, and once a distinction between arguments expressed in

---

[14] Semantic realism is often defined following Michael Dummett's definition of realism, i.e. as the thesis that statements on a given disputed class possess an objective and mind-independent truth-value. The present understanding of 'semantic realism' is better seen as a thesis concerning truth, rather than truth-value (this, of course, leaves untouched Dummett's definition, and only points to a widespread misuse of the term 'realism').

epistemic and non-epistemic terms is also allowed (cf. fn. 8), one ends up with four different versions of minimal indispensability arguments: one epistemic version for realism, one epistemic version for platonism, and their respective non-epistemic formulations (cf. Panza, Sereni, forthcoming, for more on this).

In [MIA] we identified the extra premise required for getting an argument for platonism in a premise expressing [QC]. Indeed, if one endorses [QC], and further assumes that the theories to which it is applied are true (or at least justified), platonism easily follows, as Quine's views testify. Our primary intention was that of codifying the essential structure of many current versions of IA, and it seems rather uncontroversial to do so in that way.

Different premises might be thought of, however. First of all, alternative criterion for ontological commitment can be employed.[15] Secondly, it is well possible that other sorts of considerations (e.g. related with scientific practice, or with a clarification of a proper account of the applicability of mathematics) can be made to bear here, and added to the conclusion of [MIA$_a$] as further premises in order to get a platonist conclusion.

Notice that the semantic realist who appeals to IA will face the same problem signalled by Paseu in connection with the indifference objection (cf. fn. 14): she will have to explain how a mathematical statement can be held true when no specific interpretation of its content is endorsed. But this task is by itself independent of IA. And again, the fact that IA can deliver a semantic realist conclusion without entering into details on this score might be seen as a virtue or as a vice of IA, depending on one's constraints on what an argument for realism should accomplish.

In conclusion, a clarification on Putnam's views is needed. [MIA$_a$] is, as a matter of fact, consonant with Putnam's views in the philosophy of mathematics. On the one side, it is consistent with his view on equivalent descriptions: the truth of mathematical statements can be accounted for either according to a platonist «Mathematics as Set Theory» picture, or according to an anti-platonist «Mathematics as Modal Logic» picture (cf. Putnam 1967, p. 298). On the other side, it accounts well for what Putnam took to be his original intentions in suggesting IA, i.e. «that a *prima facie* attractive position – realism with respect to the theoretical entities postulated by physics, combined with *antirealism* with respect to

---

[15] Recent attempts to reject IA on the basis of a rejection of [QC], such as e.g. that promoted by Azzouni (1998; 2004) seems to be effective only against [MIA$_b$]. Fictionalist objections to IA should be put in another basket, since fictionalists are likely to reject not only premise (*v*), but premise (*iii*) above all.

mathematical entities [...] – doesn't work» (cf. Putnam, forthcoming; cf. also Putnam 1975, p. 74; emphasis in the original). It is not easy to make Putnam's different remarks on IA across time square with each other. Putnam's aim seems always to have been primarily to defend a realist position about mathematics, not a platonist one. His adoption of Quine's criterion and his talk of acceptance of mathematical entities in the quotation expressing [PIA] reported above should be taken as a convenient way of formulating *part* of his position, but should be qualified with clarifications about both equivalent descriptions and the alleged metaphysical import of [QC].[16] If this is so, and if we leave Quine's earlier hints at IA apart, it seems fair to say that a (plausibly epistemic) version of our minimal argument for semantic realism is what gets closest to the original version of an indispensability argument as presented by Putnam himself. Nonetheless, the quotation above, i.e. [PIA], has traditionally been taken as the reference formulation of an argument for platonism. Nothing prevents it from being so taken, provided one bears in mind that doing so requires considering it as detached from the wider context of Putnam's overall views.

References

Azzouni, J. (1998): "On 'On What There is'". *Pacific Philosophical Quarterly*, 79, 1998, pp. 1-18.

Azzouni, J. (2004): *Deflating Existential Consequence. A Case for Nominalism*. Oxford-New York, Oxford University Press.

Baker, A. (2003): "The Indispensability Argument and Multiple Foundations for Mathematics". *The Philosophical Quarterly*, 53(210), pp. 49-67.

Baker, A. (2009): "Mathematical Explanation in Science". *British Journal for the Philosophy of Science*, 60, pp. 611-633.

---

[16] At the end of his (1971) Putnam complains that space is lacking to have no space for a discussion of the theory of equivalent descriptions, but adds that his own view «is that none of these approaches [i.e. descriptions] should be regarded as 'more true' than any others» (pp. 356-7). For Putnam's objections to the alleged metaphysical import of [QC], see e.g. Putnam (2004).

Batterman, R.W. (2010): "On the Explanatory Role of Mathematics in Empirical Science". *British Journal for the Philosophy of Science*, 61, pp. 1-25.

Benacerraf, P. (1965): "What Numbers Could Not Be". *The Philosophical Review*, 74, pp. 47-73; also in Benacerref and Putnam (1964), pp. 272-294.

Benacerraf, P. (1973): "Mathematical Truth". *The Journal of Philosophy,* 70, pp. 661-679, also in Benacerraf and Putnam (1964), pp. 403-420.

Benacerraf, P.; Putnam, H. (1964): *Philosophy of Mathematics. Selected Readings.* Prentice-Hall, Englewood Cliffs (N.J.); 2nd ed. 1983, Cambridge, Cambridge University Press.

Bueno, O.; Colyvan, M. (2011): "An Inferential Conception of the Application of Mathematics". *Noûs,* 45, pp. 345–74.

Colyvan, M. (2001): *The Indispensability of Mathematics.* Oxford-New-York, Oxford University Press.

Colyvan, M. (2007): "Mathematical Recreation Versus Mathematical Knowledge", in Leng *et al.* (2007), pp. 109–22.

Dieveney, P.S. (2007): "Dispensability in the Indispensability Argument". *Synthese, 157*(1), pp. 105-128.

Field, H. (1980): *Science Without Numbers.* Oxford, Blackwell.

Field, H. (1982): "Realism and Anti-Realism About Mathematics". *Philosophical Topics*, 13, 1982, pp. 45-69; also in Field (1989), pp. 53-78.

Field, H. (1989): *Realism, Mathematics and Modality.* Oxford, Blackwell.

Hellman, G. (1989): *Mathematics Without Numbers.* Oxford-New York, Oxford University Press.

Leng, M. (2002): "What's Wrong With Indispensability? (Or, The Case for Recreational Mathematics)". *Synthese* 131, pp. 395–417.

Leng, M.; Paseu, A.; Potter, M. (2007): *Mathematical Knowledge.* Oxford-New York, Oxford University Press.

Liggins, D. (2008): "Quine, Putnam, and the 'Quine-Putnam' Indispensability Argument". *Erkenntnis* 68, pp. 113–27.

Maddy, P. (1992): "Indispensability and Practice". *Journal of Philosophy*, 89, pp. 275-289.

Maddy, P. (2007): *Second Philosophy*. Oxford-New York, Oxford University Press.

Marcus, R. (2010): "The Indispensability Argument in the Philosophy of Mathematics". *The Internet Encyclopedia of Philosophy*. http://www.iep.utm.edu/indimath/http://www.iep.utm.edu/indimath/http://www.iep.utm.edu/indimath/http://www.iep.utm.edu/indimath/http://www.iep.utm.edu/indimath/http://www.iep.utm.edu/indimath/

Motterlini, M. (2002): "Reconstructing Lakatos: a Reassessment of Lakatos' Epistemological Project in the Light of the Lakatos Archive". *Studies in History and Philosophy of Science*, 33, pp. 487–509.

Panza, M.; Sereni, A. (forthcoming): *Plato's Problem. An Introduction to Mathematical Platonism*. Palgrave MacMillan, Houndmills, Basingstoke.

Parsons, C. (1978): "Quine on the Philosophy of Mathematics", in C. Parsons, *Mathematics in Philosophy*. Ithaca (NY), Cornell University Press, 1983. Also in L. Hahn, P. Schilpp (eds.), *The Philosophy of W.V. Quine*. La Salle (Ill), Open Court, 1986, pp. 369-395.

Paseu, A. (2007): "Scientific Realism", in Leng *et al* (2007), pp. 123-149.

Pincock, C. (2004): "A Revealing Flaw in Colyvan's Indispensability Argument". *Philosophy of Science*, 71, pp.61-79.

Pincock, C. (2011): "Batterman's "On the Explanatory Role of Mathematics in Empirical Science". *British Journal for the Philosophy of Science*, 62, pp. 211-217.

Psillos, S. (1999): *Scientific Realism: How Science Tracks Truth*. Oxford, Routledge.

Putnam, H. (1956): "Mathematics and the Existence of Abstract Entities". *Philosophical Studies*, 7, pp. 81-87.

Putnam, H. (1967): "Mathematics Without Foundations". *The Journal of Philosophy*, 64, pp. 5-22. Reprinted in: Putnam (1975a), pp. 43-59.

Putnam, H. (1971): *Philosophy of Logic*. Harper & Row, New York. Repr. in Putnam (1975b), ch. 20. Reference are to this latter edition.

Putnam, H. (1975a): "What is Mathematical Truth?". *Historia Mathematica*, 2, pp. 529-43, reprinted in Putnam (1975b), ch. 6.

Putnam, H. (1975b): *Mathematics, Matter and Method, Philosophical Papers vol. I*. Cambridge, Cambridge University Press (2nd ed. 1985).

Putnam, H. (1979/1994): "Philosophy of Mathematics: Why Nothing Works", in H. Putnam, *Words and Life*, ed. J. Conant. Cambridge (MA), Harvard Univeristy Press, 1994, originally published as «Philosophy of Mathematics: A Report», in *Current Research in Philosophy of Science: Proceedings of the P.S.A. Critical Research Problems Conference*, ed. P.D.Ansquith and H. Kyburg Jr., East Lansing, Philosophy of Science Association, 1979.

Putnam, H. (2004): *Ethics Without Ontology*. Cambridge (MA), Harvard University Press.

Putnam, H. (Forthcoming): "Indispensability Arguments in the Philosophy of Mathematics", in H. Putnam, *Philosophy in an Age of Science*, ed. by M. De Caro and D. Macarthur. Cambridge (MA), Harvard University Press, 2011.

Quine, W. Van O. (1948): "On What There Is". *Review of Metaphysics*, 2, pp.21-38, repr. in W.V.O. Quine, *From a Logical Point of View*. New York, Harper & Row (2nd ed. 1961), ch. 1.

Quine, W. Van O. (1986): "Reply to Charles Parsons", in L. Hahn, P. Schlipp, (eds.), *The Philosophy of W.V. Quine*. La Salle (Ill), Open Court.

Quine, W. Van O. (1995): *From Stimulus To Science*. Cambridge (MA), Cambridge University Press.

Resnik, M.D. (1995): "Scientific vs Mathematical Realism: The Indispensability Argument". *Philosophia Mathematica (III)* 3, pp. 166-174.

Resnik, M.D. (2005): "Quine and the Web of Belief", in S. Shapiro (ed.), *Oxford Handbook of Philosophy of Mathematics and Logic*. Oxford-New York, Oxford University Press, pp. 412-437.

Sober, E. (1993): "Mathematics and Indispensability". *The Philosophical Review*, 102, pp. 35-57.

Third Section
Philosophy of Psychology

Studying the Mind
On a pair of explanatory issues in cognitive science(s)

Alfredo Paternoster
University of Bergamo
alfredo.paternoster@unibg.it

1 Mechanistic explanations

In the recent discussion on foundations of cognitive sciences much attention has been paid to the notion of *mechanistic explanation* (Craver 2001; 2007). A mechanism is, roughly, a system constituted by components arranged to produce a certain goal or behavior. Mechanisms are often nested in other mechanisms, so that a full account of a mechanism requires to describe other mechanisms. For instance, the mechanistic explanation of a carburetor requires not only the description of its parts and the way these are organized to mix air and fuel, but also locating the carburetor in the context of the operation of the engine of which the carburetor is part (together with other components of the engine).

Therefore a mechanism is a hierarchically organized system and a mechanistic explanation involves intrinsically several explanatory levels. More precisely, an ideally complete mechanistic explanation describes a mechanism by integrating three perspectives (cf. Craver 2001, pp. 62-68):

(A) The "isolated" perspective describes the mechanism at its proper level; this is an ordinary causal explanation describing the input-output relations of the mechanism (level 0);

(B) The "contextual" perspective locates the mechanism in the context of some other mechanism (or mechanisms) of which it is a part and to which its activities contribute (level +1);

(C) The "constitutive" perspective breaks down the mechanism into its constitutive parts (level −1) in such a way that we can understand how these parts enable the input-output relations that are characteristic of the mechanism at the level 0.

The reader will recognize in this model a familiar kind of explanation in cognitive sciences: *functional* explanation (Cummins 1975; cf. Marraffa 2010). The emphasis on the three types of perspectives reminds Lycan's (1987) claim that the relation between function and structure is instantiated at each pair of explanatory levels. Therefore mechanistic explanations were popular in cognitive sciences even before being assessed as "mechanistic" (to put it roughly, but see *infra*, §2).

Indeed the mechanistic model of explanation seems to be the most appropriate in cognitive sciences: the characteristic multi-level structure of mechanistic explanations fits well the complexity and the variety of *explananda*.

1 Mechanisms and computations

A mechanistic model is not necessarily computational; indeed paradigmatic applications of the mechanistic model are found in (non-computational) neurosciences and in molecular biology. On the other hand, computational explanation is fundamentally mechanistic, since in computational models a function, e.g., vision, is decomposed in a collection of sub-functions (say, color detection, depth computation, ...) each realized by a specific mechanism described in algorithmic terms. Algorithms are modules that can in principle be decomposed in other modules, some of which can eventually be identified to neural mechanisms. Therefore the relation between mechanistic explanation and computationalism seems not to arise any particular problem; the question at stake is, rather, the *advisability* of having, in cognitive sciences, mechanistic models of *computational* kind.

I think, however, that successes obtained in some domains, such as vision, syntax or mental imagery, vindicate computational explanations; moreover, after about fifteen years of discussion, a large consensus grew up on the thesis that computational explanations, provided that they are not restricted to CRTM-style,[1] play an important role in cognitive sciences. We could talk about a *"liberalized"* computationalism, meaning by that that the class of eligible algorithms to compute a given function includes artificial neural networks, which actually work better in some domains. This liberalization is also reflected in the acknowledgment that what fundamentally distinguishes the different research programs and explanatory styles is the choice of the explanatory level to which restrictions or constraints on models are introduced. It has been taken for granted, in recent years, that restrictions should be applied at the neurological or more generally biological level – it is exactly this assumption that characterizes the rejection of classic computationalism. However, as Cordeschi (2002) persuasively argued, the justification of this assertion is often missing. Mechanistic models should not be interpreted as reductionist in style; on the contrary, since each level is explanatorily autonomous, mechanistic models can be regarded as an instance of *explanatory pluralism*, involving co-evolution of different disciplines; and a way to realize co-evolution consists in imposing on a given model constraints individuated both at higher and lower levels.

A good example of the relevance of high level constraints is the indispensability of a *computational theory* in Marr's technical sense (see e.g. Clark 1990). One cannot forgo, in cognitive sciences, a very high descriptive level of a cognitive process in terms of *competence*: realizing a mechanism able to provide performances similar to those of (a cognitive process of) a human agent does not amount, *per se*, to giving an explanation of that process if one lacks a high-level description of the constraints to be satisfied by a mechanism in order to be regarded as a realizer of that cognitive process. From a slightly different point of view, in absence of the competence level we would be unable, on the one hand (looking at "bottom-up"), to understand what a neurophysiological mechanism does; and on the other hand (looking at "top-down") we would probably cut the mind in the wrong slices. As Marr formerly put it, it is the computational (or competence) level which qualifies computational explanation in the first instance.

The competence level provides, on a bottom-up perspective, a re-description of common-sense *explananda*, as well as an explanatory framework (to be "filled up" with the specification of lower levels algorithms or mechanisms); and, on a top-down perspective, an interpretation of the behavior of neural mechanisms, as a system organized for a certain aim that can be individuated only at higher level.

---

[1] I am referring to the Fodorean Computational-Representational Theory of Mind, the logical-propositional model of mental processes.

For all these reasons, computational explanations still are, among the class of mechanistic explanations, prominent in cognitive sciences.

3 Horizontal *vs.* vertical expansions: the problem of integration

It is well-known that many cognitive scientists have recently insisted on the embodied and embedded nature of cognition, in some cases up to the point of removing the borders between the body (or the subject as a whole) and the environment. Dynamicist explanations, based on nonlinear dynamical systems, are familiar instances of this attitude. On this background, some authors have proposed to *integrate* mechanistic/computational explanations and dynamicist explanations.[2] However, the harmonization between the mechanistic-computationalist (from now on, M-C) style and the dynamicist one faces the following problem.

Arguably, the most plausible way of integrating dynamicist explanations and M-C models consists in designing systems whose parts (subsystems) are individuated according to a mechanistic principle; at the same time, however, since the inter-relations among parts are non-linear (for instance, they cannot be reduced to simple input/output connections), their global organization requires a dynamical description, i.e., the whole system turns out to be a dynamical system. This is what Bechtel (1998) calls an "integrated system". Integrated systems, however, are very *weakly* modular (*ibid.*), since each of their parts is influenced by the activity in some other parts of the system. Is this degree of modularity sufficient for the standard required by the M-C model? In other words, can dynamicism and modularity (to some degree) go together?

The answer is hardly positive. Carruthers (2006), for instance, argues that a M-C explanation requires constraints on the concept of part (or module) far more committing than what is required for the notion of integrated system, namely, informational encapsulation and massive modularity. Although Carruthers's argument is much controversial, it seems difficult to deny that the typical decompositional method of the M-C model is the more reliable the more the relevant mechanisms tend to be encapsulated.[3]

---

[2] The paradigm case is Clark (1997; 2008). Here I am not concerned with the case of authors who want to give up computational explanations across the board, in favour of the dynamicist models.

[3] Note that this difficulty affects, at least on one aspect, the M-C model quite independently of the issue of the integration of dynamicist explanations. Think of the

Moreover there is a further difficulty in this integration, coming from a tension between two recent tendencies in cognitive sciences, which we could call the "vertical expansion" and the "horizontal expansion". While the latter involves a view of mind as a collection of functions produced by the brain (even if the pluralist picture is not reductionist in spirit), the former downsizes the role of the brain, both ontologically and epistemologically. On the ontological side, the externalist philosophy underlying the horizontal expansion denies that mind depends ontologically from the nervous central system alone; correspondingly, on the epistemological side, explanations of mental phenomena cannot be found in the cerebral bases alone.

Is it possible to find a coherent synthesis of the two kinds of expansion? Probably yes, but provided that both sides weaken their more committing claims. More specifically, the synthesis requires a reasonable compromise in which the epistemological demands of embodiment and embeddedness are vindicated, but at least one untenable metaphysical thesis is given up. It is the idea that mind is literately *constituted* by extra-bodily items. Let us explain.

Arguing for the externalist character of mental processes explanation is correct to the extent that it is legitimate and probably necessary making use, in the description of a competence theory (a computational theory in Marr's sense), of a teleological-intentional vocabulary, which makes reference to aspects of the external environment. In order to say what is, for instance, the goal of vision, or what is computed by each of its subsystems, it is much sensible to mention environmental properties. However, this does not imply that a computational state *supervenes on* external factors, over and above the internal factors, as the metaphysical externalism pretends. In fact, a computational state can carry information about an external event or property, but this does not make that state an external "object". A representation is not identical to what it represents. In counterfactual terms, an external difference implies a mental change in an agent only if the difference is detected by (some parts of) the agent (cf. Egan 1995; Patterson 1996). It seems to me that granting this to internalism is a metaphysically moderate view that does not undermine the central point of embeddedness, that is, the influence of environment on representations.

On the other hand, there are good reasons to believe that some mental processes are characterized by the direct, "real-time" involvement of external factors. In these cases it seems correct to say that external factors are metaphysically constitutive of the relevant processes and states (cf. Clark 2008; Wilson 2004). In *this* sense the classic idea according to which

---

alleged holism of central processes, which Fodor takes long since as a threat to the entire project of explaining the mind.

cognitive processes supervene on internal cerebral states can no longer be taken for granted. Shortly stated, one can neither jump to externalism across the board, nor regard psycho-neural supervenience as a dogma.

4 Conclusions

This brief comment[4] has to be received with two *caveat*. First, never forget that the rational reconstruction of scientific models is, here like elsewhere, *idealized*. Suffice to consider that celebrated theories, such as Kosslyn's theory of mental imagery, or Johnson-Laird's mental models theory, can be subsumed under the above-discussed mechanistic model only to a certain extent. Second, despite of the explanatory virtues of some theories, our knowledge of mind remains globally poor. Just to give some examples, computational explanations seem not to be able to account for abductive reasoning (inference to best explanation); empirical evidence for models of language processing is modest; object recognition is still poorly understood. Not to mention the subject-matter of consciousness.

However, the following papers, concerning respectively language and time, the notion of extended mind, the (neuro)psychological bases of moral judgement and consciousness, give us, luckily, some reasons to be optimist for the future.

References

Bechtel W. (1998): "Representations and Cognitive Explanations: Assessing the Dynamicist Challenge in Cognitive Science". *Cognitive Science*, 22, pp. 295-318.

Carruthers, P. (2006): *The Architecture of the Mind: Massive Modularity and the Flexibility of Thought*. Oxford, Oxford University Press.

Clark, A. (1990): "Connectionism, Competence, and Explanation". *British Journal for the Philosophy of Science,* 41, pp. 195-222.

Clark, A. (1997): *Being There*. Cambridge, MIT Press.

Clark, A. (2008): *Supersizing the Mind*. Oxford, Oxford University Press.

---

[4] A larger and deeper discussion of these themes is contained in Marraffa & Paternoster (forthcoming).

Cordeschi, R. (2002): *The Rediscovery of Artificial*. Kluwer, Dordrecht.

Craver, C.F. (2001): "Role Functions, Mechanisms and Hierarchy". *Philosophy of Science*, 68, pp. 53-74.

Craver, C.F. (2007): *Explaining the Brain. Mechanisms and the Mosaic Unity of Neuroscience*. Oxford, Oxford University Press.

Cummins, R. (1975): "Functional Analysis". *The Journal of Philosophy*, 72, pp. 741-60.

Egan, F. (1995): "Computation and Content". *The Philosophical Review*, 104, 2, pp. 181-203.

Lycan, W.G. (1987): *Consciousness*. Cambridge, MIT Press.

Marraffa, M. (2010): "Funzioni, meccanismi e livelli", in Pagnini, A. (ed.), *Filosofia della medicina*. Roma, Carocci.

Marraffa, M.; Paternoster, A. (forthcoming): "Functions, Levels and Mechanisms. Explanatory Problems in Cognitive Science". *Theory and Psychology*, forthcoming.

Patterson, S. (1996): "Success-Orientation and Individualism in Marr's Theory of Vision", in K. Akins (ed.), *Perception*. Oxford, Oxford University Press.

Wilson, R. (2004): *Boundaries of the Mind. The Individual in the Fragile Sciences: Cognition*. Cambridge, Cambridge University Press.

Language and consciousness
The role of mental time travel in discourse processing

Erica Cosentino
Università della Calabria
ericacosentino@libero.it

1 Mental time travel and the self

Mental time travel (MTT) refers to the faculty of human beings to mentally project themselves backwards in time to re-live, or forwards to anticipate, events (Suddendorf and Corballis, 2007). The mental reconstruction of personal events from the past is also known as *episodic memory* in literature and has been the topic of intense research (Tulving, 1983, 2002). By contrast, the mental construction of possible events in the future has only very recently been considered. Nevertheless, several cognitive, neuropsychological, neuroimaging studies attest that projecting into the past (episodic memory) and into the future (foresight) depend on the same cognitive mechanism and neural substrate (two recent reviews are available: Suddendorf and Corballis, 2007; Schacter et al., 2008); in particular, amnesic patients, who are unable to recall past events, have been found to be equally unable to anticipate future events (Hassabis et al., 2007; Klein et al., 2002). In the light of these data, a number of investigators have recently articulated a broad view of memory in which the future-oriented function takes priority over the past. Schacter and Addis (2007) observed that episodic memory is fragmentary and fragile, and it is prone to various kinds of errors and illusions. However, these distortions don't reflect a malfunction of the memory system; on the contrary, they mirror the operation of adaptive processes. The fact that episodic memory is fragile and fragmentary suggests that its adaptive function is not related to its role as an exact record of personal history; an alternative option is that episodic memory may serve as a system to simulate future episodes. This explanation helps us to understand why memory is constructive and why it is prone to errors and alterations. Schacter and Addis' idea is that, since the future is not a repetition of the past, simulation of future events may involve a flexible system that can extract and re-combine elements of previous experiences – that is, episodic memory is a fundamentally constructive, rather than reproductive, process.

However, there would be little point in thinking about the future if the imagined future scenarios could not be translated in actions. Therefore, mental simulations about the future have to be integrated with mechanisms that motivate behavior. But how could evolution has produced a system that allows individuals to override the imperative of instant gratification of their own short-term goals? Suddendorf and Busby (2005) provide an interesting answer: the individual identifies himself with the future self and makes the imaginary future self's goal his own. In this way, individuals can secure not just the present, but also future survival. This means that MTT may fulfil its adaptive function as a sophisticated and flexible system of motive priority management by creating a *temporally extended self*. This hyphotesis about the origin of the extended self accounts also for common sense intuition that individuals' personal identity is equally composed by both their memories and their future plans. Neuropsychological literature confirms this hypothesis by reporting on a lot of cases in which a damage to the neural area responsible for self-projection in time (mainly hippocampus and prefrontal cortex) causes the loss of the sense of self (Schacter, 1996). In spite of this, some authors are critical about the neurocognitive reliability of the notion of the self. They think that the self is a social construction, acquired by means of the most important human cultural artifact: language. I propose that, in order to clarify the relationship between language and the self, we should consider the link between language and MTT.

In the following sections, I consider the role of MTT in language from two points of view. The first one is adopted by cognitive linguistics, according to which language is rooted in cognition; in particular, following this approach I point out that temporal language depends on MTT. This is a very important outcome for the general case for the involvement of MTT in language functioning. However, the specific case for MTT involvement in discourse processing requires more than such a claim. In order to clarify this topic, I adopt the cognitive pragmatics point of view, according to which communication involves the construction of a balance between a speaker and a listener. As humans communicate with other predominantly utilizing a distinctly story-like structure, I would like to show that the pragmatic balance between a speaker and a listener is sustained by a cognitive brain network that underlies discourse processes. Therefore, in order to account for the role of MTT in discourse processes, one has not to consider mental time travel in isolation, but its relation with other mental components.

2 Time and language

In order to analyze the relationship between language and MTT, a crucial step entails analyzing how organisms build their event representations. The main point here is that the representation of an event depends on the temporal perspective of the individual constructing it; in other words the individual's temporal perspective constrains his representation of the event. Drawing a parallel with an individual's spatial projection may help us to clarify the point. Changing spatial perspective implies the ability to understand that the way in which an object appears depends on our point of view; for example, the way the room in which we find ourselves appears depends on where we are seated: though we represent our environment in a certain way, we know that the same place would appear differently just by changing our position in space. Based on the perceptual information available at the time, we can imagine the same reality as it might appear from different vantage points in space. The same is true for the temporal equivalent, which implies understanding that the way an individual thinks of events depends on his own perspective in time: just as moving through space we see the same objects differently, moving in time events are seen differently. Understanding that we can have multiple temporal perspectives for the same event implies, for example, that we have the ability to think of a future event as something that at some point we will find in the past. By changing the temporal perspective we change the way we represent the event but also the temporal relationship between that event and others, in other words changing perspective has systematic effects.

Representing events is fundamental to our ability to speak about them and since MTT is a necessary component of this representation it is also a necessary component of language. More specifically, encoding the relationship between the event and an organism's temporal perspective is carried out by temporal language, principally by tense. Tense indicates the relationship between two elements: the time in which a sentence is spoken (speech time) and the time of the reported event (event time). The tense system implies a third component beyond the time of an event and the time of speaking of the event: it implies 'reference time' (Reichenbach, 1947).

As to the crucial relevance of MTT for language, McCormack & Hoerl (1999), having carried out an analysis of the components of temporal language, maintain that the concept of 'reference time' depends on grasping the fact that different times offer alternative temporal perspectives on events. What they define as *temporal perspective taking*, that is the ability to decentralize one's own temporal point of view and flexibly take on an alternative perspective, is the basis of reference time. Temporal perspective taking is centred on episodic memory but also involves taking on a flexible future temporal perspective, so we can say that this function is carried out by MTT. An analysis of reference time's involvement in temporal language acquisition could furnish the degree of involvement of MTT in the makeup of cognitive content of linguistic expressions. Studies of linguistic development in children have shown that MTT is not involved in the acquisition of the first tenses used by children, which according to McCormack & Hoerl might depend on *script* acquisition; reference time is in effect a component of mature temporal language but is absent in the initial stages of development – coherent with later acquisition of MTT (Grant & Suddendorf, 2009).

These considerations overall allow us to achieve an important result: language ability (following cognitive linguistics' approach) rests on cognitive systems, therefore it cannot be independent of cognition. In particular, temporal language depends on MTT. Insofar as this is a significant outcome, it is nonetheless not enough to account for the central question that concerns us here, that is how to account for speakers'ability to comprehend and product discourse. The central theme here is (following cognitive pragmatics' approach) the speakers' capacity to create a communicative balance by the interaction between several mental components, in particular those that form the self-projection system.

3 Self-projection and the brain

Recently, Buckner and Carroll (2007) investigated the relationship between MTT, mindreading and some forms of navigation in space. They pointed out that there is a close functional and structural relationship between these three elements. In fact, despite the adaptive characteristics that distinguish each component from the other two, they rely on a common set of processes called by Buckner and Carroll "self-projection". Traditionally, the diverse abilities that depend on self-projection have been considered individually, but, according to Buckner and Carroll, there is a shared brain network that supports them, which includes frontal and medial temporal systems.

By using self-projection, humans can build an imaginary state of affairs, that is, a simulation of a mental scenario. I would like to propose that the ability to build alternative scenarios by employing self-projection abilities may be the key to understand how the discourse processing level works. For example, when human beings build a narration about an event, they have to detach themselves from the present and be able to mentally explore alternative spatial, temporal and social perspectives. During a conversation, a speaker has to be able to adopt the perspective of the interlocutor in order to understand his communicative aims, but he has also to track spatial and temporal informations that frame the conversation. Therefore, discourse processes entail several forms of self-projection in time, space and other's mind. Are there any empirical evidence for this claim?

A preliminary consideration is that we should distinguish between *macrolinguistic* analysis' level (concerning pragmatic and discourse level processing), as opposed to a *microlinguistic* one (concerning lexical and morpho-syntactic skills) (Davis et al., 1997). In several cases, the study of pathological language has revealed that macro and microlinguistic abilities can be dissociated; in particular, the microlinguistic dimension may be intact while the macrolinguistic one is damaged. From these damages, we can learn a lot about the involvement of self-projection in discourse comprehension and production.

The strong link between mindreading and pragmatics has already been deeply investigated. In particular, literature about autism shows that if mindreading system is impaired and, subsequently, people are unable to recognize other individuals' mental states (such as intentions), then linguistic ability of autistic people is also impaired. More to the point, autistic individuals' linguistic deficit doesn't concern syntax or semantics, but pragmatics. In particular, testing autistic people's narrative abilities using a picture-description task, it emerged that discourse level processing is a relative weakness, mainly when they had to narrate stories in which social instead of mechanical intentionality should have been reconstructed (Fletcher et al., 1995).

Recently, the same correlation between self-projection brain and narrative language has been reported also in the case of the spatial component. Individuals affected by Williams Syndrome have a strong visual-spatial deficit, in particular with reference to the capacity to reorient one's self in one's environment (Lakusta et al., 2010). When tested on a picture-description task, they showed good phonological, lexical and syntactic skills, but their story descriptions were less effective than those produced by the typical language development group on measures assessing global coherence and lexical informativeness, showing dissociation between macro and microlinguistic abilities (Marini et al., 2010). It should be noted that, at the present time, it isn't yet clear if Williams narrative deficit is caused by their visuo-spatial deficit, however this interpretation can't be ruled out and it seems not unreasonable to think of Williams narrative deficit in these terms.

At last, I would like to focus my attention on the relationship between MTT and macrolinguistic abilities.

4 Mental time travel and discourse level processing

Consider a situation in which at some point in a conversation a doubt arises as to what is being said, perhaps because we don't know if the speaker is being ironic or not. When there is such a doubt our consciousness notes that something isn't working and, in doing so, the automatic information processing systems give way to conscious processes which we can summarize as those processes at the basis of an implicit question of the 'I understand you correctly?' variety. In such cases, in order to re-establish a convergence between speaker and listener we must mentally travel back in time to re-analyze the conversation. The listener has to dissociate himself from the present and flexibly take on a view from the past (let's keep in mind that mental projection to the future might just as well be involved).

The re-balancing of speaker and listener requires checking and monitoring the conformity between the self and the other; this process is carried out making the conversation's time axis explicit. The conformity check implies in fact a self-monitoring through time: the individual flexibly assumes alternative temporal perspectives and summarizes them in an integral glance in virtue of the knowledge that these perspectives all belong to the same self, that is in virtue of the knowledge of one's own extension in time. The role carried out by MTT is even more clear when considering the discourse production plane, where MTT's functioning is particularly weighty.

At the level of discourse production the conformity check between the self and the other serves to plan and manage one's own discourse while keeping track of the other's knowledge. In these terms MTT is also a check mechanism of one's own inter-temporal coherence and coordination through time. An exemplary case in seeing these processes at work is that of 'influencing' in which the speaker wants to produce a change in the mental states of his listeners. Influencing indeed doesn't only require the reading of minds (those of the listeners – what others think – or one's own – what the speaker wishes the listener to think), but working through the structure of the discourse in order to produce a change. The relevant issue here is therefore the self-monitoring of the speaker to verify if he is supplying all the information needed to be understood. The question which guides the interpretative process is centred not on the other but on the self: 'Am I making myself understandable?'. Conformity checking arises therefore in negotiating between different times in order to balance flexibility need – set by the changing external situation – with the perseverance need set by the goal (therefore being coherent).

In summary, mental time travel is the instrument that we use to represent the temporal dimension of discourse as long as the self in time is the bond that unites the various points or various temporal perspectives along a common axis. The validity of this explanation could be verified in analyzing what happens in the situation in which it is not possible to rest on the temporal framework that MTT gives to conversation. Schizophrenia is an example of an extreme case in which communicative equilibrium is broken and temporal framework is missing.

5 Schizophrenia

Many meta-analyses and reviews of cognitive deficits in schizophrenia have consistently shown that episodic memory is particularly compromised (Danion et al., 2007; Neuman et al., 2007; Danion & Huron, 2007; Toulopoulou & Murray, 2004; Aleman, Hijman, de Haan et al., 1999). Recent studies have verified that not only the ability to recall past events is hampered but also the ability to imagine future ones: therefore schizophrenia could specifically involve a deficit of MTT (D'Argembeau, Raffard, Van der Linden, 2008; Danion, unpublished results). Herein I am suggesting that damage to MTT is at the origin of schizophrenic linguistic deficit.

An initial observation is that the language of schizophrenics is not compromised at the grammar level nor is it on the plane of meaning but rather it is compromised on the discourse level (Andreasen, Hoffman & Grove, 1985; Marini et al., 2008). In particular, in the study of Marini and colleagues (2008), the linguistic assessment was performed on story-telling and they found that language production in schizophrenia is impaired mainly at the macrolinguistic level of processing. The symptoms associated with an abnormal language ability in schizophrenics are vast; the most common one is, according to Andreasen's (1979) famous classification, 'derailment', that is the loss of a conversation's goal in gradual steps. Other commonly observed symptoms include 'loss of goal', 'tangentiality' (indirect or irrelevant answers) and 'lack of content' (vagueness of words and answers which provide reduced information).

All of these manifestations have a key linguistic symptom in common which, according to McGrath (1991), is the 'lack of planning and execution'. It is this problem which compromises ability in many other areas apart from language functions and it affects non verbal communication as well (Frith 1992). According to Chaika (1990; see also Covington et al., 2005), at the linguistic level this results in a 'loss of voluntary control on the organization of discourse'. According to Frith's (1992) proposal, difficulties in keeping track of other's knowledge in order to manage one's own discourse projects are at the basis of language disorders in schizophrenics. The problem therefore principally involves programs for production. This claim has been confirmed in an experimental study in which volunteers had to describe a coloured disc so as to let the listener know which disc to choose from between other coloured discs (Cohen, 1976); the results are very interesting for us; in fact it was noted that communication failed only when the description was made by the schizophrenic patient, who, on the other hand, had no problem in performing the task on the basis of the normal subject's description. It would seem then that normal subjects do not understand schizophrenics while schizophrenics do indeed understand normal subjects.

In these patients language represents a paradigmatic case for the observation of what happens when the conformity check cannot be performed. Schizophrenics' language is characterised by a dis-phasing of the two processes of comprehension and production. The conformity check fails because in the process of discourse construction the patient is unable to keep track of the self in the negotiations between the self and the other. Nonetheless, the problem cannot be traced back to a simple malfunctioning of self-awareness. Self-awareness is a basic condition in triggering checks and monitoring of conformity between the self and the other; in fact MTT includes self-awareness as one of its basic components. However this component does not indicate which path to follow in order to solve the conformity check problem: self-awareness alone cannot carry out this function. The time factor should be emphasised here, that is the ability to reflect upon the temporal dimension of discourse; the central aspect being discussed here is in fact monitoring the self over time and negotiating between different times, which is possible only on such a basis. In sum, schizophrenics' macrolinguistic deficit seems to validate the idea of the role of temporal self-projection in discourse level processing. Looking at this conclusion in the light of previous considerations about the relation between several forms of self-projection and discourse processes leads us to interesting final remarks.

6 Conclusions

I have proposed that discourse comprehension and production may be understood as a form of scenario building that enables mental exploration of perspectives and events beyond those that emerge from the immediate environment. This scenario building is made possible by means of some self-projection abilities like mental time travel, mindreading and spatial navigation, which share a common brain network. Self-projection in time, space and other's mind may be selectively impaired respectively in schizophrenia, Williams Syndrome and autism. These malfunctions are all characterized by a distinctive dissociation between micro and macrolinguistic abilities, with discourse level processing strongly damaged. Therefore, the study of linguistic deficits in autism, Williams Syndrome and schizophrenia proves that self-projection underlies discourse processing; in fact when one of three self-projection brain's components is impaired, individual's ability to elaborate discourse is threatened. The linguistic deficit is selective in two ways: first, with reference to other linguistic levels of analysis; second, with reference to the specific self-projection brain's damaged component. These data supports the claim that discourse comprehension and production involves the neural network of the self-projection brain. This conclusion leads us to a more general observation.

Despite language and consciousness have been dissociated according to the cognitive science traditional view of mind, if we turn our attention from microlinguistic (grammar and meaning) to macrolinguistic dimension (narrative and discourse) it emerges that the classic approach to language as a single syntactic system cannot account for linguistic processes at the discourse level processing. The role of some self-related functions in linguistic processes has to lead to a new extended view of the language nature.

References

Aleman, A.; Hijman, R.; de Haan, E.H.F.; and Kahn, R.S. (1999): "Memory Impairment in Schizophrenia: a Meta-Analysis". *American Journal of Psychiatry*, 156, 9, pp. 1358-1366.

Andreasen, N.C. (1979): "Thought, Language and Communication Disorders: 2 Diagnostic Significance". *Archives of General Psychiatry*, 36, pp. 1325-1330.

Andreasen, N.C.; Hoffman, R.E.; and Grove, W.M. (1985): "Mapping Abnormalities in Language and Cognition", in A. Alpert (ed.), *Controversies in Schizophrenia: Changes and Constancies*. New York, Guildford Press, pp. 199-226.

Buckner, R.L.; Carroll, D. C. (2007): "Self-Projection and the Brain". *Trends Cogn. Sci.*, 11, pp. 49-57.

Chaika, E.O. (1990): *Understanding Psychotic Speech: Beyond Freud and Chomsky*. Springfield, Charles C. Thomas.

Cohen, B.D. (1976): "Referent Communication in Schizophrenia: The Perseverative Chaining Model". *Annals of the New York Academy of Science*, 270, pp. 124-141.

Covington, M.A.; He, C.; Brown, C.; Naçi, L.; McClain, J.T.; Fjordbak, B.S.; Semple, J.; and Brown, J. (2005): "Schizophrenia and the Structure of Language: The Linguist's View". *Schizophrenia Research*, 77, pp. 85-98.

D'Argembeau, A.; Raffard, S. and Van der Linden, M. (2008): "Remembering the Past and Imagining the Future in Schizophrenia". *Journal of Abnormal Psychology*, 117, pp. 247-251.

Danion, J.-M. (unpublished results). *Remembering the Past and Envisioning the Future: An Investigation in an Investigation in Schizophrenia*.

Danion, J.-M.; Huron, C. (2007): "Can We Study Subjective Experiences Objectively? First-Person Perspective Approaches and Impaired Subjective States of Awareness in Schizophrenia." in Zelado, P.D.; Moscovitch, M. and Thompson E. (eds.), *The Cambridge Handbook of Consciousness*. Cambridge (GB), Cambridge University Press, pp. 481-498.

Danion, J.M.; Huron, C.; Vidailhet, P.; Berna, F. (2007): "Functional Mechanisms of Episodic Memory Impairment in Schizophrenia". *The Canadian Journal of Psychiatry*, 52, 11, pp. 693-701.

Davis, A.G.; O'Neil-Pirozzi, T.; and Coon, M. (1997): "Referential Cohesion and Logical Coherence of Narration After Right Hemisphere Stroke". *Brain and Language*, 56, pp. 183-210.

Fletcher, P.C.; Happé, F.; Frith, U.; Baker, S.C.; Dolan, R.J.; Frackowiak, R.S.J. *et al.* (1995): "Other Minds in the Brain: A Functional Imaging Study of "Theory of Mind" in Story Comprehension". *Cognition*, 57, pp. 109-128.

Grant, J.B.; Suddendorf, T. (2009): "Preschoolers Begin to Differentiate the Times of Events From Throughout the Lifespan". *European Journal of Development Psychology*, 6, pp. 746-762.

Hassabis, D.; Kumaran, D.; Vann, S.D.; and Maguire, E.A. (2007): "Patients With Hippocampal Amnesia Can Not Imagine New Experiences". *Proc. Natl Acad. Sci. USA*, 104, pp. 1726-1731.

Klein S.B.; Loftus J.; and Kihlstrom J.F. (2002): "Memory and Temporal Experience: The Effects of Episodic Memory Loss on an Amnesiac Patient's Ability to Remember the Past and Imagine the Future". *Social Cognition*, 20, pp. 353- 379.

Lakusta, L.; Dessalegn, B.; and Landau, B. (2010): "Impaired Geometric Reorientation Caused by Genetic Defect". *PNAS*, 16, 107, 7, pp. 2813-2817.

Marini, A.; Martelli, S.; Gagliardi, C.; Fabbro, F.; and Borgatti, R. (2010): "Neuropsychological Correlates of Narrative Language in Williams Sindrome". *Journal of Neurolinguistics*, 23, 2, pp. 97-111.

Marini, A.; Spoletini, I.; Rubino, I.A.; Ciuffa, M.; Banfi, G.; Siracusano, A.; Bria, P.; Caltagirone, C. and G. Spalletta (2008): "The Language of Schizophrenia: An Analysis of Micro- and Macrolinguistic Abilities and their Neuropsychological Correlates". *Schizophrenia Research*, 105, pp. 144-155.

McCormack, T.; Hoerl, C. (1999): "Memory and Temporal Perspective: the Role of Temporal Frameworks in Memory Development". *Developmental Review*, 19, pp. 154-182.

McGrath, J. (1991): "Ordering Thoughts on Thought Disorder". *British Journal of Psychiatry*, 158, pp. 307-316.

Neumann, A.; Philippot, P.; and Danion, J.-M. (2007): "Impairment of Autonoetic Awareness for Emotional Events in Schizophrenia". *Canadian Journal of Psychiatry*, 52, pp. 450-456.

Reichenbach, H. (1947): *Elements of symbolic logic*. New York, Macmillan.

Schacter, D.L. (1996): *Searching for memory. The Brain, the Mind, and the Past*. New York, Basic Books.

Schacter, D.L.; Addis, D.R. (2007): "The Cognitive Neuroscience of Constructive Memory: Remembering the Past and Imagining the Future". *Philosophical Transactions of the Royal Society (B) Biological Sciences*, 362, 1481, pp. 773-86.

Schacter, D.L.; Addis, D.R.; and Buckner, R.L. (2008): "Episodic Simulation of Future Events: Concepts, Data, and Applications". *Ann. NY Acad. Sci.*, 1124, pp. 39-60.

Spalletta, G.F.; Spoletini, I.; Cherubini, A.; Rubino, I.A.; Siracusano, A.; Piras, F.; Caltagirone, C. and Marini, A. (2010): "Cortico-Subcortical Underpinnings of Narrative Processing Impairment in Schizophrenia". *Psychiatry Research: Neuroimaging*, 182, 1, pp. 77-80.

Suddendorf, T. (2006): "Foresight and Evolution of the Human Mind., *Science*, 312, 5776, pp. 1006-1007.

Suddendorf, T.; Busby, J. (2005): "Making Decisions with the Future in Mind: Developmental and comparative identification of mental time travel". *Learning and Motivation*, 36, pp. 110-125.

Suddendorf, T.; Corballis, M. C. (2007): "The Evolution of Foresight: What is Mental Time Travel and is it Unique to Humans?". *Behav. Brain Sci.*, 30, pp. 299-313.

Toulopoulou, T.; Murray, R.M. (2004): "Verbal Memory Deficit in Patients with Schizophrenia: an Important Future Target for Treatment". *Expert Rev Neurother*, 4, 1, pp. 43-52.

Tulving, E. (1983): *Elements of Episodic Memory*. Oxford, Clarendon Press.

Tulving, E. (2002): "Episodic Memory: From Mind to Brain". *Annual Review of Psychology*, 53, pp. 1-25.

The notion of "boundary"
and the Extended Mind paradigm

Barbara Giolito
Vita-Salute San Raffaele University, Milan
barbara_giolito@libero.it

1 Personal Mind and Extended Mind: the notion of "boundary"

The so-called "Extended Mind" model, a research paradigm concerning the nature of the mind theorized by Clark and Chalmers (1998), supposes that the mind is spread outside the body boundaries. Fundamental for such paradigm is the idea – proposed by functionalism – that the causal role played by an item makes it a *mental* item: elements external to the body, that could be defined "mental" in case they would be realized by the body, should be defined "mental". The Extended Mind model seems to work for the most impersonal cognitive processes, but it seems to face difficulties in case we try to extend such a model to the subjective processes: the Extended Mind paradigm seems unsuitable for the so-called "personal mind", that is the model of the mind theorized by the Folk Psychology.

The concept of "boundary" seems to be an important key-concept in the Extended Mind paradigm: the question concerning the extension of the mind outside the body can be reinterpreted in the terms of the question concerning the possibility to not reduce the mind boundaries to the body boundaries. But the concept of "boundary" is a notion not easily applicable: it seems to be a vague and problematic notion. As regard to this, Varzi (2005) proposed some interesting considerations. Although boundaries are in some cases well determined entities, sometimes they are not well defined entities: in many cases, boundaries are not visible because they are not part of the *real* entities. Such a fact suggests a distinction as to the notion of "boundary": the distinction between "natural boundaries" (or *de re* boundaries) and "artificial boundaries" (or *de dicto* boundaries). But such a distinction between *de re* boundaries and *de dicto* boundaries, and the plausibility of the notion of "*de re* boundary", are not unproblematic: the hypothesis to consider boundaries, as well-defined, real entities rises some problems, consisting in the difficulties to specify the relations between an object and its boundaries. The relation between an object and its boundaries seems to involve a reciprocal dependence: the definition of an object depends on the boundaries of that object, but the object boundaries depend on the object itself. Such considerations make the existence of real *de re* boundaries doubtful. Varzi suggests that all boundaries are, at least in part, *de dicto* boundaries; however, such interpretation doesn't imply that boundaries are arbitrary entities; rather, they could have pragmatic bases (the human daily customs could determine the object delimitations, and the usefulness of such delimitations could make a boundary better than another).

The hypothesis to consider boundaries as entities determined by pragmatic decisions can be applied to the analysis of the mind. The notion of "mind" is hardly definable *per se*: it could be possible to raise doubts about the possibility to consider the mind boundaries as exact and definite entities. The mind boundaries could be not determined by *de re* properties of the related cognitive processes: the decision concerning the opportunity to consider a phenomenon as *mental* could depend on pragmatic considerations based on the utility of such decision. Such a point of view doesn't imply a reinterpretation of the notion of "mind" in arbitrary terms: the possibility to determine the mind boundaries in a certain way could depend on the fact that such an interpretation is acceptable in the socially shared explanation of the cognitive processes. The relevant question could not concern the hypothesis to enclose the mind inside the body, but, rather, the hypothesis according to which folk psychological concepts are useful and essential. In particular, we could identify as cognitively relevant only the cognitive processes – internal or external to the body – *causally* relevant for the explanation of the human behaviour, or we could consider as relevant also the Folk Psychology intentional terms.

2 Folk Psychology: the difficulties of Behaviourism and Simulation Theories

The Folk Psychology is the basis for most of human relations: we explicate human behaviour in terms of beliefs, desires and so on.

Nevertheless, some important attempts to deny the Folk Psychology validity has been pursued in the past; the so-called "Behaviourism" is a typical example of such attempts. Behaviourism proposes the idea that the notion of "mental state" plays no role in the explanation of the mental phenomena, because mental states are unobservable and mysterious entities: the psychological explanation should use notions related to observable entities, such as the *behaviour* of subjects involved in the psychological analysis. The behaviour observation appeared insufficient to explain the complex psychological life of human beings (for example, as suggested by Chomsky (1968), it seems impossible to explain human language by means of the behaviour of the human speakers): as a consequence, Behaviourism has been largely abandoned. Nevertheless, the general idea to deny the validity of the Folk Psychology notions has not been abandoned: such an idea is at the basis of the so-called "Simulation Theories". The Simulation Theories propose the hypothesis that the analysis of the notion of "mental life" suggested by the Folk Psychology is not based on principles concerning the way the human mind works: on the contrary, the Folk Psychology explanation of mental life would be based on the human capacity to *simulate* the behaviour of other human beings. The Simulation Theories suppose that human beings can assume the point of view of other human beings to imagine what they would do: in such a way, human beings can expect the behaviour of other human beings by attributing to them actions that they would perform. The explanation of such a process in terms of *attribution of mental states* is just descriptive and derives from the *simulation* ability: the explanatory role is played by the *simulation*.

It is possible to distinguish at least two different positions referred to as "Simulation Theories": the *Radical Simulation Theory* and the *Moderate Simulation Theory*. The Radical Simulation Theory maintains that psychological explanations must give up any introspective move; on the contrary, the Moderate Simulation Theory doesn't make such a request. Gordon (1986) is one of the most important proponents of the Radical Simulation Theory. In Gordon's opinion, the psychological explanation must be based on two processes: the *mental simulation* and the *rising strategy*. As to the *simulation*, Gordon suggests the idea that we need to assume the point of view of other human beings to understand and predict their psychological life: we imagine to be the person whose behaviour we are trying to understand, and we try to imagine our own reactions to attribute same reactions to that person. Psychology concerns reality, it doesn't concern mental states as such. Imagine to consider the following question: "do you believe that x?". To give an answer to such a question you should take into account, not the belief concerning x, but the content of "x". You evaluate "x" and, as a consequence (for example, if you assert that "x"), you place "I believe that" before "x" to obtain "I believe that x". In such a way, you can obtain the semantic level concerning mental states. The so-called *rising strategy* consists in such a move, and can be equally well applied to mental states belonging to other human beings. Suppose to consider the following question: "does Gloria believe that x?". To formulate an answer for such a question, you can imagine to be Gloria and you can try to imagine what is the answer you could give to the same question from her point of view. If your answer from Gloria's point of view is "x", you put "Gloria beliefs that" before "x" to obtain "Gloria beliefs that x". In Gordon's opinion, the essential questions concerning psychological states don't concern *mental* states; they concern *real* states. Such an explanation seems to work for particular mental states, as *beliefs*. Nevertheless, it seems to rise some problems in case of other mental states: when we try to generalize the rising strategy to other mental states, it seems we are no more able to distinguish different kinds of mental states. For example, consider *desires* (a fundamental mental state category in Folk Psychology): the content of "x" by itself seems to be not enough to differentiate the "*desire* that x" from the "*belief* that x". It seems we need another distinction to differentiate the "*desire* that x" from the "*belief* that x": the distinction between different *mental attitudes* concerning the content of "x".

Although the Radical Simulation Theory faces some problems, the Simulation Theory could be a valid psychological explanation in its moderate version. Alvin Goldman is one of the most important proponent of the so-called "Moderate Simulation Theory" (see on this Goldman 1989 and 2006). Differently from Gordon, Goldman admits a role for mental elements in the simulation process. We observe (or we imagine to observe) the behaviour of the other human beings to anticipate their real behaviour: we imagine to put ourselves from the point of view of other human beings to imagine our own behaviour from that point of view and to attribute the same behaviours to other human beings. In some cases, we don't need mental elements to realize such moves; but in most of complex cases, we need mental concepts. For example, suppose that Gloria is lost in a forest. Gloria could be enthusiastic for such intriguing situation, and she could courageously try to find her way-home. But Gloria could also be frightened and desperate, and her reaction could be an irrepressible crying. To evaluate what is the most likely reaction of Gloria, we need concepts such as Gloria's *fear* concerning wild animals, Gloria's *anxiety* in solitude case, and so on: such concepts are mental concepts. Goldman admits that we need some mental elements to understand and anticipate some behaviours of the other human beings: but the psychological states we need are, in Goldman's opinion, not the mental states of those human beings, but *our own* mental states. During the simulation, we analyse our own mental states: we observe our own possible reactions and we anticipate reactions of the other human beings by attributing to them our possible reactions. We obtain knowledge of our mental states by means of *quasi-perceptual* modalities (by means of phenomenal or neural elements): theoretical or conceptual modalities don't play any role in the psychological explanation. Nevertheless, also the Moderate Simulation Theory rises some problems and seems to need a sort of reintroduction of the Folk Psychology mental concepts. In particular, simulation can be usefully used to analyse the content of mental states to predict the related behaviours, but the *understanding* and the *explanation* of a mental state (that is, the understanding and the explanation of what *is* a particular mental state) needs the Folk Psychology *functional role* of that mental state (see Meini 2007).

3 The relevance of the Folk Psychology: a difficulty for the Extended Mind

The phenomena typically explained by means of the Folk Psychology concepts seem to be not explicable by means of the Radical Simulation Theory notions. The Moderate Simulation Theory can avoid some of the problems of the Radical Simulation Theory, but such a step seems to be achieved by means of a sort of acceptation of the Folk Psychology concepts. Both Behaviourism and Simulation Theories seem to be unable to eliminate the Folk Psychology concepts. The Folk Psychology concepts are the mental concepts typically used by human beings during their daily life: when we analyse the actions performed by other human beings, at least in the more complex situations, we need to consider such human beings as *persons*, that is entity characterised by mental states.

We previously proposed the idea that the debate concerning the Extended Mind paradigm must be analysed not in terms of "mind boundaries", because of the vagueness of the concept of "boundary". On the contrary, the debate concerning the Extended Mind paradigm could be reformulated in terms of an analysis of the possibility to reject the Folk Psychology concepts, that is, the concepts typically implying an interpretation of human beings as *persons*. As suggested by Michele Di Francesco (2007), the most relevant problem for the Extended Mind paradigm consists in its incapacity to take into account the notion of "person". The personal mind is the locus of subjectivity and rationality, it makes reference to a subjective ontology and designs a *subjective space* - characterized by the first person point of view and expressing an individual perspective – which requires intentional language in order to be described. Personal mind makes us able to explain human *actions*: it characterises the *space of reasons* with its normative and intentional features. Mere causal informational connections – characterizing cognitive systems in the Extended Mind paradigm – are not sufficient to explain the kind of unity essential to our notion of *personhood* and *subjectivity*. External cognitive processing has no phenomenological content: many forms of cognitive processing are not accessible to consciousness – they don't imply any *it is like to be* – but, to be considered as mental, they should be strictly related to conscious processing. For example, as regard to the sub-personal level operating in the brain, the outputs of brain-realized cognitive processing become accessible to our conscious mind in a more intimate and direct way than the output processed by external devices. There is a distinction between *sub-personal* and *non-personal* cognitive processing: both are external with regard to personal mind, but they entertain a different relation to it. For instance, sub-personal processing – but not non-personal processing – exhibits immunity from error through misidentification. The *mineness*, the unity of the mind and the *perspectivalness* of experience characterize the personal mind and produce a gap between personal and extended mind. Moreover, the connection between the external device contents and the mental contents is *causal*, and not *motivational*: it doesn't explain the acts of a human being as *actions*. The external device contents should be assimilated to the personal mind of a human being in order to be considered as *reasons*. Rationality, normativity and phenomenology are constitutive aspects of subjectivity: if we renounce them, we also renounce the notion of *subject*.

As previously suggested, the validity of the Extended Mind paradigm could depend on the possibility to eliminate, from the psychological analysis, the analysis of the concepts related to the notion of "person", that is, the concepts typically considered in Folk Psychology; but rationality, normativity and phenomenology do not extend themselves outside personal mind. The analysis of the Simulation Theories gives an example of some difficulties related to the attempt to reject the Folk Psychology concepts: such difficulties suggest the hypothesis that Folk Psychology concepts are not eliminable. The Extended Mind paradigm is not able to take into account such concepts: as a consequence, it seems to be unsuitable to represent an *exhaustive explanation* of the human mind. The Extended Mind paradigm could be useful to explain some aspects of the cognitive faculties; nevertheless, as long as it is not able to take into account the Folk Psychology concepts (and in case we don't have any plausible alternative to the Folk Psychology concepts), the Extended Mind paradigm seems to be unable to represent the whole explanation of the mental life.

References

Chomsky, N. (1968): *Language and Mind*. New York, Harcourt Brace Jovanovich.

Clark, A.; Chalmers, D. (1998): "The Extended Mind". *Analysis*, 58:1, pp. 10-23.

Di Francesco, M. (2007): *"Extended Cognition and the Unity of Mind. Why We Are Not 'Spread into the World'",* in De Caro, A.; Ferretti, F. and Marraffa M. (eds.), *A Cartography of the Mind*. Dordrecht, Kluwer, pp. 213-227.

Goldman, A. (1989): "Interpretation Psychologized". *Mind and Language*, 4, pp. 161-185.

Goldman, A. (2006): *Simulating Mind*. Oxford, Oxford University Press.

Gordon, R. (1986): "Folk Psychology as Simulation*"*. *Mind and Language*, 1, pp. 158-171.

Meini, C. (2007): *Psicologi per natura. Introduzione ai meccanismi cognitivi alla base della psicologia ingenua*. Roma, Carocci.

Varzi, A.C. (2005): "Teoria e pratica dei confini*"*. *Sistemi Intelligenti*, 17:3, pp. 339-418.

On the nature of moral conflicts

Maria Grazia Rossi
University of Messina
mgrazia.rossi@gmail.com

1 Emotions and moral judgment

Humans are capable of moral judgment. They have extremely sophisticated beliefs about how they should or should not behave in morally significant situations. What is the nature of these moral choices? To answer this question, we shall examine the study of those cognitive devices which provide the basis for the processing of moral beliefs and, in particular, we will focus on the idea that the process of formation of such beliefs depends heavily on the action of emotional mechanisms. We will argue that emotions play such a role because of two properties that characterize their nature. Emotions have an essential evaluative character and have a strong homeostatic role: they are evaluative systems underlying the equilibrium relationship between organism and environment (see on this Damasio, 1994).

Most scholars who address the relationship between emotions and moral judgments are primarily interested in the idea that moral judgments are essentially based on intuitions. In particular, against the rationalist models based on the supremacy of justifications (e.g. Piaget, 1932; Kohlberg, 1987), the argument is that justifications are, in most cases, only *post hoc* rationalizations and that in order to explain the process of moral judgment formation, it is necessary instead to examine moral intuitions (types of evaluations, produced by rapid, automatic and unconscious processes: see Haidt, 2001; Hauser, 2006; Hauser *et al.*, 2007). Emotions have an extremely important role in this process. We will defend this argument by focusing on the evaluative character of emotions.

From our point of view the reference to emotions can also be used to highlight a more general property of moral choice: its conflicting nature. From observations on the proto-moral behaviour of bonobos and chimpanzees, de Waal (1997 and 2006) points out that to be moral animals means firstly reaching an equilibrium between individual interests (the needs of individuals) and social motivations (within their own group and between different groups) as a part of a continuous cycle of conflicts and reconciliations. Our argument is that this dynamics of compromises can be used to emphasize the extent of the conflict that characterizes, to varying degrees, also any judgment of moral appropriateness. Using data from the study of moral dilemmas, we will show that reaching such a compromise depends on a competition process between processing systems within the mind (see also Cushman and Greene, 2011). More specifically, our goal is to argue that emotions function as an *orchestrating mechanism* in managing the relationship and internal equilibrium between elaboration systems (Ferretti, 2007; Tooby and Cosmides, 2008). To do this, we will examine the conflict between deontological and consequentialist judgments through the analysis of different psychological processes underlying them and provide evidence in support of the idea that where there is a conflict between judgments (or behaviours) in conflict, the activity of an orchestrating system becomes necessary (see Greene *et al.*, 2004 and 2008).

2 Constraining intuitions

The idea that moral judgments are largely the product of forms of rational and conscious deliberation was, in the second half of the last century, the dominant viewpoint in moral psychology. In conformity with this idea, Piaget (1932) argued in favour of a strong parallelism (in the genesis and in actual functioning) between logical and moral thinking, focusing in particular on the dependence of the second on the first and arguing that morality should be considered a logic of action. From this point of view, moral judgment is the product of a logically correct form of reasoning which is based on rules that justify the issuance of that judgment.

However the pioneering works conducted in recent years by Jonathan Haidt and his collaborators show that to explain the formation of moral judgments is not enough to make reference reasoning and reflection (see Haidt, 2001; Haidt and Hersh, 2001; Wheatley and Haidt, 2005). According to Haidt, when we assess moral issues, we almost never behave like a judge who seeks the truth by objectively considering the arguments and evidence.

Instead, we are lawyers: we already have a perspective from which to start and then we proceed in search of evidence that might persuade others of the truth of our opinion (Haidt, 2001). The issue of consensual incest illustrates well the point in question:

> Julie and Mark are sister and brother. They are traveling together in France on summer vacation from college. One night they are staying alone in a cabin near the beach. They decide that it would be interesting and fun if they tried making love. At the very least, it would be a new experience for each of them. Julie is already taking birth control pills, but Mark uses a condom, too, just to be safe. They both enjoy making love, but decide not to do it again. They keep that night as a special secret, which makes them feel even closer to each other. What do you think about that? Was it OK for them to make love? (Haidt, 2001, p. 814)

Most of the people interviewed by Haidt consider as morally unacceptable the behaviour of Mark and Julie. The relevant fact is that those same people are unable to justify why they believe this is the case. With the term *moral dumbfounding*, Haidt refers to the inability of subjects to justify their moral convictions. This is just one aspect. When asked to explain, the opinions of subjects, far from being based on wholly inadequate grounds, seem extremely strong and entrenched. This is noticeable when, faced with explanations that stress the idea that incestual sex leads to genetic abnormalities in the offspring or damages the relationship between brother and sister, the experimenter notes that the arguments are unfounded because in the context of the scenario it is observed that such experience has made their relationship stronger and that the brother and sister had used at least two forms of birth control. Faced with replies of this type from the experimenters, only 17% of the subjects changed their original opinion. All the other respondents, instead of proceeding with a revision of their judgments, continue to argue that it is wrong even if they are unable to explain why (see Haidt, 2001 and 2006; Haidt and Hersh 2001).

The metaphor of the lawyer and the cases of *moral dumbfounding* show that, contrary to that maintained in rationalist models, a greater understanding of moral competence implies a shift of attention away from moral reasoning to the study of moral intuitions. Haidt and Bjorklund write:

> "Moral intuition" is defined as the sudden appearance in consciousness, or the fringe of consciousness, of an evaluative feeling (like-dislike, good-bad) about the character of actions of a person, without any conscious awareness of having gone through steps of search, weighing evidence, or inferring a conclusion. [...] This conscious experience of blame or plaice, including a belief in the rightness or wrongness of the act, is the moral judgment. (Haidt and Bjorklund, 2008, p. 188)

The argument is that moral intuitions are laden with emotions. The reference to David Hume is obvious. In his Treatise of Human Nature the Scottish philosopher wrote that "morality, therefore, is more properly felt than judg'd of" (Hume, 1739, p. 930). Moral judgments are based on this emotional feeling.

Emotions can play a role like this because they have a strong evaluative character. Assigning such a property to emotions means acknowledging to them a task in the assignment of an emotional marking at certain features (of objects or of events) that may have particular biological-evolutionary importance for the organism (Damasio, 1994). By virtue of this characteristic, emotions cause moral judgments, motivate our choices and preferences; moral reasoning, at least in these cases, is only a *post hoc* construction that struggles to find reasons to justify already expressed moral verdicts.

2.1. Emotional prejudices

If moral judgments are strongly driven by emotional intuitive judgments then it would suffice to manipulate the emotional component to influence the moral beliefs of subjects. Experimental evidence supports a conclusion of this type. Wheatley and Haidt (2005) show that the induction in hypnotised subject of a brief pang of disgust when reading a particular word (in the experiment half of the subjects are induced to feel disgust when they read the word *take* and half when they read the word *often*) is enough to provoke, in the assessment of the stories actually containing those words, harsher judgments of moral inappropriateness. It is even more surprising that the subjects use emotionally induced information to judge a case in which no violation is present. Subjects are presented with a text like this:

> Dan is a student council representative at his school. This semester he is in charge of scheduling discussions about academic issues. He [tries to take/often picks] topics that appeal to both professors and students in order to stimulate discussion. (Wheatley and Haidt, 2005, p. 782)

Subjects condemn Dan when one of the two words (*take* or *often*), laden with disgust, is present in the text even though, paradoxically, the only "immoral" act that the student may have committed is to have encouraged greater discussion between teachers and students. In this regard, the justifications provided by some of the subjects are illuminating, highlighting well the *post hoc* nature of moral reasoning: Dan is to some a snob looking for popularity, while several subjects dismiss any kind of explanation and allow themselves to be dominated by their feeling of disgust exclaiming phrases such as: "It just seems so weird and disgusting" (Wheatley and Haidt, 2005, p. 783). The authors of the experiment argue that the feeling of disgust linked exclusively to an arbitrary word is used unconsciously by the subjects during the assessment of the stories as information about the immorality of the act in question (see Schnall *et al.*, 2008).

3 Moral conflicts

By highlighting the character of evaluation, we have emphasized the idea that emotional intuitions are a driving force of moral judgments. There is no doubt that emotions accompany and influence our moral judgments; in moral psychology, the dispute focuses rather on the temporal and causal role attributed to emotional mechanisms (see Huebner *et al.*, 2009). Despite this, evidence of a philosophical, psychological and neuroscientific order support the hypothesis that emotional devices are at least one of the mechanisms underpinning the way in which moral judgments are formed (see Greene and Haidt, 2002; Prinz, 2007; Sinnott-Armstrong *et al.*, 2010 and Young *et al.* 2010).

However, from an evolutionary point of view, according to de Waal (2006), this feature of emotions must be taken in connection with a more general characteristic that defines moral choice. This should be considered the result of a compromise between social habits of cooperation and competition that are in conflict with each other. A form of behaviour or a judgment that is morally appropriate always calls into question, to different degrees, an equilibrium between individual interests and social motivations.

The study of cognitive devices underpinning the processing of moral dilemmas can be incorporated into such an interpretative framework. In fact, moral dilemmas impose an assessment between competing interests, between concomitant and conflicting obligations because they are based on opposite reasons. Hypothetically, we could present a number of good reasons for both the possible actions but when one is faced to a dilemma one cannot do both of them. We are condemned to moral failure. Our idea is that the conflict that underlies the paradox of the decision we are called to make when we are faced to a dilemma is the reflection of an internal conflict between competing elaboration systems. The difficulty, as shown by Cushman *et al.* (2010) is to explain how the different systems come into play in such conflicting relationship.

In a series of studies, Joshua D. Greene stressed this point by testing empirically the hypothesis that traditional definitions of consequentialism and deontology are also good descriptions of natural psychological types:

the philosophical equivalent of two distinct psychological systems (see Greene, 2008; Greene *et al.* 2002; 2004 and 2008). In particular, Greene point out:

> deontological judgments tend to be driven by emotional responses and that deontological philosophy, rather than being grounded in moral reasoning is to a large extent an exercise in moral rationalization. This is in contrast to consequentialism, which [...] arises from rather different psychological processes, ones that are more "cognitive," and more likely to involve genuine moral reasoning. (Greene, 2008, p. 36)

Deontological and consequentialist judgments would thus be connected with two types of different psychological processes. What happens when the conflict between the two possible options is more urging? The *crying baby dilemma* exemplifies the fact of such matter:

> It's wartime, and you and some of your fellow villagers are hiding from enemy soldiers in a basement. Your baby starts to cry, and you cover your baby's mouth to block the sound. If you remove your hand, your baby will cry loudly, the soldiers will hear, and they will find you and the others and kill everyone they find, including you and your baby. If you do not remove your hand, your baby will smother to death. Is it okay to smother your baby to death in order to save yourself and the other villagers? (Greene, 2008, p. 44)

When the participants of the study evaluate similar scenarios, the alternative representations of the two possible options generate a conflict between the two underlying psychological systems: the activity of the brain regions devoted to the management of cognitive conflicts (in particular the anterior cingulate cortex) increases as well as those involved in the processes of abstract reasoning and cognitive control (dorsolateral prefrontal cortex); in particular the response times increase when people demonstrate a preference for the consequentialist option, thus judging the violation as permissible. Against the hypothesis that this increase is simply proportional to the complexity of the cost-benefit analysis, Greene defends the thesis that it is rather an external indication of a conflict between the cognitive and emotional systems. Since the emotional response makes us inclined not to accept as permissible a moral breach (don't smother the baby!), in order to reach a consequentialist verdict (The baby will die no matter what; save yourself and the others), the activity of the emotional system must be nullified and overcome. To perform such an operation, the cognitive system needs time (see Greene *et al.*, 2008). In these cases, to formulate a consequentialist judgment, we must somehow go against that emotional feeling that, otherwise, would be responsible for the production of deontological judgments.

4 The orchestration function of emotions

Starting from the study of cognitive devices underlying moral conflicts, we have discussed the idea that these dilemmas are the result of a conflict within the mind, between two types of different elaboration systems. An *orchestrating mechanisms* becomes necessary precisely where there is a conflict between judgments (or behaviours) in conflict. In these cases, what is at stake is moral equilibrium, that is, the appropriateness of our judgments. Our hypothesis is that emotions are, as orchestrating mechanisms, the guarantors of such an equilibrium.

Designed by evolution to produce automatic and immediate responses, emotional devices allow in emergency situations, by motivating an appropriate course of action, the management of events which are particularly relevant to the survival of the organism. Damasio writes:

> In other words, the biological "purpose" of the emotions is clear, and emotions are not a dispensable luxury. Emotions are curious adaptations that are part and parcel of the machinery with which organisms regulate survival. Old as emotions are in evolution, they are a fairly high-level component of the mechanisms of life regulation. You should imagine this component as sandwiched between the basic survival kit (e.g., regulation of metabolism; simple reflexes; motivations; biology of pain and pleasure) and the devices of high reason, but still very much a part of the hierarchy of life-regulation devices. For less-complicated species than humans, and for absentminded humans as well, emotions actually produce quite reasonable behaviors from the point of view of survival. (Damasio, 1999, pp. 73-74)

From this point of view, emotions play largely a homeostatic role and it is as a result of this that emotions can be integrated within the broader class of regulatory mechanisms: systems delegated to the management of the restoring of the functional equilibrium between organism and environment. The interplay between the breaking and restoration of functional equilibrium allows the consideration of the usefulness of emotional devices in the management of the equilibrium between organism and environment (see Craig, 2002 and 2003; Damasio, 1994; Frijda, 1986; Plutchik, 1994). Now, from the perspective of cognitive mechanisms that enable effective appropriate behaviour, consistent with that maintained by Ferretti (2007, pp. 81-86), the problem of external equilibrium needs to be addressed taking into account what is, first and foremost, a equilibrium within the mind. In our opinion the homeostatic function can be exploited in this sense to propose a hypothesis where emotions play a crucial role. Emotions manage the relationship and internal equilibrium between processing system.

In this regard, the comments by Tooby and Cosmides seem particularly interesting (see Tooby and Cosmides, 1990 and 2008; Cosmides and Tooby, 2000). Starting with considerations on the cognitive architecture of the human mind, the two authors thematize precisely the need for an orchestrating system:

> In general, to behave functionally according to evolutionary standards, the mind's many subprograms need to be orchestrated so that their joint product at any given time is coordinated to deal with the adaptive challenge being faced, rather than operating in a self-defeating, discoordinated, and cacophonous fashion. We argue that such coordination is accomplished by a special class of programs: the emotions that evolved to solve these superordinate demands. In this view, the best way to understand what the emotions are, what they do, and how they operate is to recognize that mechanism orchestration is the function that defines the emotions, and explains in detail their design features. (Tooby and Cosmides, 2008, p. 117)

The emotional devices that evolved to recognize meaningful recurrent situations, through their orchestrating function would thus be suitable to activate a specific configuration of sub-programs appropriate to that specific circumstance. Our idea is that emotions function as an orchestrating mechanism also in the management of the—competition between the processing systems underpinning moral dilemmas.

Yet, at first sight, for example if we stop to think about *the crying baby dilemma* that we discussed in the previous paragraph, the plausibility of this hypothesis does not seem so evident and warranted. Instead, we are led to reconsider the idea of emotions not as a *mechanism of orchestration* but rather of emotions as a *mechanism of interference*. Essentially, one could continue, the emotional system simply interferes, hinders and often even takes precedence over the cognitive system. From this viewpoint, emotions would function rather as a mechanism of interference and all that could be argued is that, at least in these cases, despite the presumed rationality of the consequentialist option, we simply continue to prefer the emotional intuitions. The example of *moral dumbfounding* discussed in paragraph 2.1 could also be interpreted in a similar manner: in what sense should a judgment made passively, following the induction of disgust in hypnotized subjects, also have value from a moral perspective? The interviewees consider the behaviour of Dan inappropriate only because of the feeling of disgust which, though an element entirely alien to the moral level, interferes in evaluation. What evidence do we have to avoid this conclusion?

An important indication in support of our hypothesis comes from pathology. A study by Koenings *et al.* (2007) examines the opinions of a group of subjects with bilateral lesions to the ventromedial prefrontal cortex (VMPC group). The cognitive profile of patients with brain damage such as these is characterized by emotional deficits (both in emotional responses and in emotional regulation) but with preserved ability of general intelligence, logical reasoning and awareness of moral and social norms. In studying the responses of this group, two control groups are linked: one composed of neurologically normal subjects (neurologically normal comparison, NC) and one composed of patients with brain lesions of different types (brain-damaged comparison, BDC). In high conflict moral dilemmas, the percentage of consequentialist judgments increases dramatically for individuals belonging to the VMPC group. This result allows the authors to conclude that when judgments require the evaluation of competing considerations as in the case of high conflict dilemmas, emotions play a crucial role.[1] When emotional devices do not function properly, the consequentialist alternative becomes the default option and moral choice tends to coincide with a mere cost-benefit analysis. The moral appropriateness is skipped.

5 Conclusions

In discussing a number of cases of *moral dumbfounding*, we have used the dissociation between judgments and justifications to push to the core of the matter concerning the nature of moral judgments, intuitive, fast and automatic mechanisms and to defend the idea that, thanks to their evaluative nature, emotional intuitions should underpin the formation of such judgments.

Starting from the premise that the moral choice has a conflicting nature, we have then maintained that it is possible to assign to emotions also a more general function. In this regard, considering moral dilemmas as a case study, we have put forward a dual hypothesis: (1) the external conflict underpinning moral dilemmas is a reflection of an internal conflict between competing processing systems, (2) emotions function as an *orchestrating mechanism* in the management of the relationship and internal equilibrium between processing systems.

Upon reaching such conclusion, which identifies in the dimension of the conflict a key feature of moral choice, the appropriateness of our

---

[1] Also the experimental results of Ciaramelli et al. (2007) are congruent with this reading.

155

judgments and our moral equilibrium is therefore devolved on the function of emotional orchestration.

References

Ciaramelli, E.; Muccioli, M.; Làdavas, E.; and di Pellegrino, G. (2007): "Selective Deficit in Personal Moral Judgment Following Damage to Ventromedial Prefrontal Cortex". *Social Cognitive and Affective Neuroscience*, 2, pp. 84-92.

Cosmides, L.; Tooby, J. (2000): "Evolutionary Psychology and the Emotions," in Lewis, M. and Haviland-Jones, J.M. (eds.), *Handbook of Emotions*, 2nd Edition. New York, Guilford, pp. 91-115.

Craig, A.D. (2002): "How do You Feel? Interoception: the Sense of the Physiological Condition of the Body". *Nature Reviews Neuroscience*, 3, pp. 655-666.

Craig, A.D. (2003): "A New View of Pain as a Homeostatic Emotion". *Trends in Neurosciences.* 28, pp. 303-307.

Cushman, F.A.; Greene, J.D. (2011): "Finding faults: How Moral Dilemmas Reveal Cognitive Structure", in Decety, J. and Cacioppo, J. (eds.), *The Handbook of Social Neuroscience*. New York, Oxford University Press.

Cushman, F.; Young, L.; and Greene, J.D. (2010): "Our multi-System Moral Psychology: Towards a Consensus View", in Doris, J.; Harman G.; Nichols, S.; Prinz, J.; Sinnott-Armstrong, W. and Stich, S. (eds.), *The Oxford Handbook of Moral Psychology*. Oxford, Oxford University Press, pp. 47-71.

Damasio, A.R. (1994): *Descartes' Error: Emotion, Reason, and the Human Brain*. London, Penguin.

Damasio, A.R. (1999): *The Feeling of What Happens*: *Body and Emotion in the Making of Consciousness*. New York, Harcourt Brace and Co.

Ferretti, F. (2007): *Perché non siamo speciali*. Bari-Roma, Laterza.

Frijda, N.H. (1986): *The Emotions*. Paris, Editions de la Maison des Sciences de l'Homme, Cambridge University Press.

Greene, J.D. (2008): "The Secret Joke of Kant's Soul", in Sinnott-Armostrong, W. (ed.), *Moral Psychology, Vol. 3: The Neuroscience of Morality: Emotion, Disease, and Development*. Cambridge (MA), MIT Press, pp. 35-79.

Greene, J.D.; Haidt, J. (2002): "How (and Where) Does Moral Judgment Work?". *Trends in Cognitive Sciences*, 6, 12, pp. 517-523.

Greene, J.D.; Nystrom, L.E.; Engell, A.D.; Darley, J.M.; and Cohen, J.D. (2004): "The Neural Bases of Cognitive Conflict and Control in Moral Judgment". *Neuron*, 44, pp. 389-400.

Greene, J.D.; Morelli, S.A.; Lowenberg, K.; Nystrom, L.E.; and Cohen, J.D. (2008): "Cognitive Load Selectively Interferes With Utilitarian Moral Judgment". *Cognition*, 107, 3, pp. 1144-1154.

Haidt, J. (2006): *The Happiness Hypothesis: Finding Modern Truth in Ancient Wisdom*. New York, Basic Books.

Haidt, J. (2001): "The Emotional Dog and Its Rational Tail: A Social Intuitionist Approach to Moral Judgment". *Psychological Review*, 108, 4, pp. 814-834.

Haidt, J.; Biorklund, F. (2008): "Social Intuitionist Answer Six Questions about Moral Psychology," in W. Sinnott-Armostrong (ed.), *Moral Psychology, Vol. 2: The Cognitive Science of Morality: Intuition and Diversity*. Cambridge (MA), MIT Press, pp. 181-217.

Haidt, J.; Hersh, M. (2001): "Sexual Morality: The Cultures and Reasons of Liberals and Conservatives". *Journal of Applied Social Psychology*, 31, pp. 191-221.

Hauser, M.D. (2006): *Moral Minds: How Nature Designed Our Universal Sense of Right and Wrong*. New York, Ecco Press/Harper Collins.

Hauser, M.D.; Cushman, F.; Young, L.; Jin, R.K. and Mikhail, J. (2007): "A Dissociation Between Moral Judgments and Justifications". *Mind & Language*, 22, 1, pp. 1-21.

Huebner, B; Dwyer, S.; and Hauser, M. (2009): "The Role of Emotion in Moral Psychology". *Trends in Cognitive Sciences*, 13, 1, pp. 1-6.

Hume, D. (1739): *A Treatise of Human Nature: Being an Attempt to introduce the expermental Method of Reasoning into Moral Subject.* London, Thomas Longman (quoted from *Trattato sulla natura umana*, It. trans. by P. Guglielmoni, Milano, Bompiani, 2001).

Koenings, M.; Young, L.; Adolphs, R.; Tranel, D.; Cushman, F.; Hauser, M.; and Damasio, A. (2007): "Damage to the Prefrontal Cortex Increases Utilitarian Moral Judgments". *Nature*, 446, pp. 908-911.

Kohlberg, L. (1984): *The Psychology of Moral Development: Moral Stages and the Life Cycle.* San Francisco, Harper & Row.

Piaget, J. (1932): *Le jugement moral chez l'enfant.* Paris, F. Alcan.

Plutchik, R. (1994): *The Psychology and Biology of Emotion.* New York, HarperCollins.

Prinz, J. (2007): *The Emotional Construction of Morals.* New York, Oxford University Press.

Schnall, S.; Haidt, J.; Clore, G.L.; and Jordan, A.H. (2008): "Disgust as Embodied Moral Judgment". *Personality and Social Psychology Bulletin*, 34, 8, pp. 1096-1109.

Sinnott-Armstrong, W.; Young, L.; Cushman, F. (2010): "Moral Intuitions," in Doris, J.; Harman, G.; Nichols, S.; Prinz, J.; Sinnott-Armstrong, W. and Stich, S. (eds.), *The Oxford Handbook of Moral Psychology.* Oxford, Oxford University Press, pp. 246-272.

Tooby, J.; Cosmides, L. (1990): "The Past Explains the Present: Emotional Adaptations and the Structure of Ancestral Environments". *Ethology and Sociobiology*, 11, pp. 375-424.

Tooby, J.; Cosmides, L. (2008): "The Evolutionary Psychology of the Emotions and their Relationship to Internal Regulatory Variables," in Lewis, M.; Haviland-Jones J.M. and L.F. Barrett (eds.), *Handbook of Emotions, 3rd Edition.* New York, Guilford, pp. 114-137.

de Waal, F. (1997): *Good Natured: The Origins of Right and Wrong in Humans and Other Animals.* Cambridge (Mass), Harvard University Press.

de Waal, F. (2006): *Primates and Philosophers*. Princeton (NJ), Princeton University Press.

Wheatley, T.; Haidt, J. (2005): "Hypnotic Disgust Makes Moral Judgments More Severe". *Psychological Science*, 16, 10, pp. 780-784.

Young, L.; Bechara, A.; Tranel, D.; Damasio, H.; Hauser, M.; and Damasio, A. (2010): "Damage to Ventromedial Prefrontal Cortex Impairs Judgment of Harmful Intent". *Neuron*, 65, 6, pp. 845-851.

The Embodied Nature of Self-Experience

Maria Francesca Palermo
Roma Tre University
palermo.mf@gmail.com

1 How the notion of body schema and body image explains phenomenal consciousness

We will analyze the issue of human self-consciousness by adopting an *ecological approach*. Such a hypothesis assumes that the identification of mental states and experiences is partly carried out on the basis of the phenomenal properties of conscious experience. According to this view, each conscious experience is experienced from a *first person perspective* and conveys a primitive, pre-linguistic and preconceived form of self-consciousness (Bermúdez et al. 1995).

By emphasizing the embodied nature of experience, our main goal is to attribute a key role to the qualitative experience of actions in terms of both the cognizance of the identification of the other, as well as the self-ascription of psychological states. The emergence of a "bodily model of the self" is based on a sub-personal and automatic process through which different perceptual and sensory aspects are tied together so as to produce a "coherent product". Psychometric studies have recently demonstrated that the sense of having a body is made up of various sub-components, among which, *the sense of ownership, the sense of agency, and the sense of action* (Haggard 2005; Blakemore & Frith 2003). Other important components which the conscious self is anchored upon include visceral sensations and basic emotions. The experience of being an "incarnate self" is quite simply a holistic construction of the subject characterized by part-whole relationships

161

and derived from very diverse sensory sources. With regards to this, neuropsychology has hypothesized the existence of a *body schema* (Head & Holmes 1911) that is an unconscious brain map which is continuously updated in relation to the position of limbs as well as to shape and physical posture. The body schema is a system of sensory-motor processes that function without reflective awareness or the necessity of perceptual monitoring. The proprioceptive information deriving from kinesthetic, visual, and somatic sources, as well as vestibular and balance functions, all contribute to structuring the body schema. This, indeed, provides the basis for sensory self-consciousness. Essentially, from a phenomenological perspective, consciousness is thus not only "incorporated", but it is also characterized by a primitive form of self-reference which depends on a sense of possession and agency. These forms of experience are currently topics of great interest in the quest to identify specific neural correlates through empirical research (Chaminade & Decety 2003). Furthermore, recent neuropsychological data suggest that we must distinguish between the notion of body schema and the perceptive and conceptual awareness of the body as an object, which forms the basis of the identification of the *body image*, or "the system of perceptions, attitudes, beliefs, and dispositions pertaining to one's own body" (Gallagher 2005, p. 37). In relation to neuropsychology, the distinction between body image and body schema is, in fact, supported by the identification of a double dissociation. It's is quite possible that the functions of body schema and of body image are selectively damaged, thus leading to specific problems concerning consciousness and subjectivity. In very extreme cases of *deafferentation*, the functioning of the body schema is compromised since the patients in such cases are almost completely lacking in tactile and proprioceptive experience, and must put their trust in an extremely articulate body image (Gallagher & Cole 1995). Cases of *unilateral spatial negligence*, on the other hand, constitute a radical alteration of body image which yet leaves the functions of the body schema completely unaltered (Berlucchi & Aglioti 1997). According to Melzack (1997) corporeal awareness relies upon a large neural network where somatosensory cortex, posterior parietal lobe and insular cortex play crucial and different roles in the tactile and proprioceptive deficits, but there is no evidence they can cause alterations of *higher-order* body awareness.

Furthermore, neuroscientific data indicate that each successful extension of the *behavioral space* is reflected in the neural substratum of the body image. One important aspect concerning these studies regards the evolution of the use of *substrata* by primates (Maravita & Iriki 2004). Some recent studies have shown that Japanese macaques can be trained to use tools, although it is quite rare for them to use these tools under natural conditions. When a macaque uses a tool, changes in specific neural networks of its brain take place, and this discovery suggests that the tool is temporarily integrated in the animal's body schema. When the macaque is given a piece of food that is placed out of its reach and the animal can use a rake to draw the food nearer, a change in the macaque's physical self-model can be observed in the animal's brain. As a matter of fact, it is as though the model of its hand and the space surrounding it were extended to the tip of the tool. The brain constitutes an internalized image of the tool, which is thus assimilated to the existing body image. The flexibility of the body schema seems to depend on the properties of the body maps that are codified in the parietal lobe. These results point out that "the body schema is not only sensory-motor but also action-oriented in nature […;] these findings indicate that body schema is characterized both by multi-sensory integrations and dynamic plasticity" (Gallese & Sinigaglia 2010, p. 747).

The key step in human development, therefore, may have been that of rendering a larger part of conscious experience entirely available or, to be more precise, accessible to consciousness. As soon as we are able to consciously experience our *body in action* (as in the case of a tool being used) we are able to pay attention to this process, optimize it, establish concepts regarding it, and control it by performing authentic volitions. This sense of the body in action makes the attribution of sense of agency and sense of ownership possible, thus conceiving "the bodily self as an integrated system characterized by matching of sensory-motor integration" (Legrand 2006, p. 108). This investigation will allow us to specifically characterize a bodily self, rooted in sensori-motricity.

2 Mentalization and development of the self

Conscious self-experience is clearly a gradual phenomenon because it grows in intensity as the organism's own sensitivity regarding an internal context increases and as the organism extends its self-control skills towards the external environment. There seems to be a unitary principle that subtends the aforementioned discoveries: some of the strata of our self-model serve as bridges to a social dimension since they can directly map the abstract internal descriptions of what happens within us to the descriptions of what happens in others. These processes support an implicit and physical idea of *intersubjectivity*.

It seems that our comprehension of others' minds would thus be founded upon an *access* that is primarily *introspective* and which is built upon a system of low-level *sensory-motor resonance* which draws from motivational and emotional resources of self-comprehension as well as comprehension of others (Gallese & Goldman 1998; Jeannerod & Pacherie 2004). This idea was already present a century ago under the concept of *empathy* (Lipps 1903) and expresses the possibility we have to examine the modes of understanding behaviors and social activity. However, being able to feel what individuals feel, to put oneself in their place, can also be problematical. More recently the idea of *mental simulation* has played an essential role in the philosophy of mind debate. A complex interdisciplinary dispute on the nature of folk-psychology has been going on between *Theory theorists* (Gopnik & Meltzoff 1997; Stich & Nichols 1992) and *Simulation theorists* (Goldman 1995; Gordon 1995; Harris 1995). While there are many varieties and different views of *Simulation Theory* (ST), all have in common that *simulation acts* are a very effective device for forming predictions and explanations. The basic idea of simulation theory, in fact, is that one uses one's own mental experience as internal model to simulate other's mind and thereby comes to predictions and explanations. The process itself is structured as an internal representational simulation. One motivation for simulation is that while the prediction of other people's behavior can be difficult, the prediction of "our own immediate and near immediate actions is usually a simple and accurate matter. We thus set up an internal figure" (Davies & Stone 1995, p. 15). From this perspective, the attribution of mental states is not totally based on inferential mechanisms or conceptualization processes. In other words, simulating means recruiting "the same perception and action systems that are called upon during interaction" (Gallagher & Jeannerod 2002, p. 21). Due to these characteristics, it is often defined as *off-line* simulation (Meini 2007) since it indicates the automatic, subconscious and sensory-motor activation of a series of neural mechanisms triggered by the observation of others' behavior. In off-line simulation, one takes one's own *decision-making system* off-line and supplies it with pretend inputs of beliefs and desires of the person one wishes to simulate in order to predict their behavior. Alvin Goldman's (1995; 2006) view of simulation theory is a very interesting account of simulation. According to Goldman, the fundamental resource of interpretation interacts with knowledge and generalizations (ToM). In accordance with Goldman's so-called "moderate" position, we feel it plausible that evolution connected much more than just one development strategy in self-identification and in agency ascription. In his important 1989 paper "Interpretation Psychologized" in *Mind and Language*, he does not propose that simulation is the only method used for interpersonal mental ascriptions or for prediction of behavior, but simulation is *nothing less* than an initial comprehension method in the minds of others, yet were this to fail, the use of a theoretical framework could be a plausible strategy for both phylogenetic and ontogenetic development. While we know that we sometimes use simulation to predict other people's behavior, we are typically not aware of simulation processes going on in us. Goldman suggests that this is because simulation need not be an introspectively vivid affair and because it is likely that the process is "semi-automatic, with relatively little salient phenomenology" (Goldman 1995, p. 88). This view fits in well with the evidence gained through research about *autism* and *mirror neurons*. Autistic people, who are generally quite bad with imagination and especially with pretend play, show an impairment of the mindreading ability. Baron-Choen formulates his thesis quite succinctly, and his formulations make clear its relationship to simulationism:

> Empathy involves a leap of imaginations into someone's head. While you can try to figure out another person's thoughts and feelings by reading their face, their voice and their posture, ultimately their internal world is not transparent, and in order to climb inside someone's head one must imagine what it is like to be them (cf. in Goldman, 2006, p. 201).

Simulation theorists, who see a connection between the reduced ability of pretend play and mindreading, suggest that autism is good evidence that we normally use mental simulation. Furthermore, the discovery of *mirror neurons* in the early 90's has provided a solid demonstration of basic simulative mechanisms, thus supporting the simulation theory. The basic idea is that the mirror system can be a precursor of the more general skill of *mind-reading*. It has been suggested that these neurons are used for *imitation*, allow the *acquisition of language* (Rizzolatti & Arbib 1998) and enable *theory of mind* (Gallese & Goldman 1998). Gallese and Goldman suggest that mind-reading could make a contribution to inclusive fitness since "detecting another agent's goals and/or inner states can be useful to an observer because it helps him anticipate the agent's future actions, which might be cooperative, non-cooperative, or even threatening" (Gallese & Goldman 1998, pp. 495-496). Mirror neurons facilitate the creation of pretend (*off-line*) action (*motor images*) that correspond to the visually perceived actions of others. They also claim that simulation can be used to retrofit as well as predict mental states. In other words, it is possible to determine which mental states of a target have already occurred. The concept of mental simulation has been expanded to show how some neurons have multimodal *mirror* properties: in other words, they represent both the motor and sensory nature of action. As such, these *multimodal mental representations* constitute an essential element in the interaction that takes place between agents and the environment, as well as in *interpersonal* relationships.

These representations are part of a system which simulates the information required for movement and furthermore, allows for it to be felt or sensed even without external stimulation. Simulating, indeed, means re-assembling an internal replication of an event, reactivating, at least in part, the neuronal activation *patterns* tied to the experience of that event. According to simulation theorists, mirror neurons help us to translate our visual perception of the other person's behavior into a mental plan of that behavior in ourselves, thus enabling an explanation or prediction of the other person's thoughts or action. This is an ecological approach to the minimal sense of *bodily self*, in which action and non-conceptual qualitative proprieties of experience comes in as the basis for *grasping* the self-world distinction. It is clear that any simulation requires perceptual *information* and neural activation. This embodied mechanism can be used to generate mental explanations of the target's behavior and is used to bootstrap such as understanding. The similarity of activated areas between observation of action, mental simulation and imitation accounts for a shared neural representations of bodily knowledge. This is the first step in the direction of genuine understanding of the pre-reflective consciousness.

Thus, our thesis is that a complete account of substantive self-consciousness will have to deal with the issues raised by the structural representation of bodily knowledge. One way to summarize these pre-theoretical condition is that understanding of others in terms of their mental states requires a "massively hermeneutic background [...] This kind of knowledge derives from *embodied practices* in second-person interactions with others" (Gallagher 2001, p. 84). Our ability to represent another's thoughts are intimately tied together and may have similar origins within the brain. Thus it, can be proposed that the human system involved in the perception and understanding of actions also requires identification with the other, as well as the capacity to distinguish the self from other selves. We believe that all these mechanism are necessary to experience intersubjectivity. After all, in the next section we examine the first cognitive mechanisms that gives subjects an understanding of oneself and others as persons, that is to say the ability to see the self and other as two distinct members in the category of agents.

2 Imitation, simulation and first interactions

Young children are capable of understanding and perceiving others in terms of embodied interactions. Mental states that are in essence private to the self may be shared between individuals. Experimental studies have documented that starting from the very first weeks of life infants are able to progressively imitate a number of gestures observed, for instance protrusion of the tongue, eyebrow movements, rotation of the head, finger movements, and, above all, gestural characteristics that are useful for expressing primary emotions like surprise, joy, boredom, and general vocalizations (Kugiumutzakis 1998). The studies mainly confirm the ability on the part of newborns to exhibit finely attuned inter-coordination of the movements and facial expressions of the adults they are in contact with. As such, this ability, which is defined as *primary intersubjectivity* (Trevarthen 1979) would shed new light on *imitative re-enaction* and on newborn learning, as well as on the stages moving from embodied simulation of actions to simulation of the mind. In other words, it would appear that primary forms of intersubjectivity cannot be solely conceived as mechanisms that are antecedent to the development of an authentic theory of mind. Rather, they constitute the basis of our *interpersonal relationships* and hold a key role in all stages and interactions of life.

The studies of infant imitation suggest that the infant has both a primitive body schema and some degree of proprioceptive performative awareness. These are mechanisms that operate as general conditions of possibility for motor stability and control but are also directly related to the possibility of imitation. To imitate, the infant must perceive another's acts and use this as a basis for an action plan with its own body. This translation must be accomplished without verbal instruction. As a matter of fact, according to simulation theory, a child simulates the body of an adult not as a mere reflexive mechanism, but though a process which Meltzoff and Moore (1998; 2002) define as *"active intermodal mapping"* (AIM) which in turn defines a *"supramodal actual space"* (SAS). The resonance mechanism of the motor system fits well the Meltzoff-Moore theory: the emotional link between observed action and executed action is because the personal perspective of the demonstrator is linked to the interior and physical perspective of the child. This occurs through a specific phenomenal state of intentional consciousness (Gallese 2005). Experiments by Meltzoff and Moore demonstrate that from birth the action of the infant and the perceived action of the other person are coded in the same "mental code": a *cross-modal system* that is directly attuned to the actions and gestures of other humans. The translation is already accomplished at the level of innate body-schema that integrates sensory and motor systems (Gallagher & Meltzoff 1996). In the human infants this *intermodal system* accounts for the possibility of recognizing and imitating other humans. One interesting consequence of this notion of *supramodality* is that there is a primordial connection between *self* and *other*. This innate capacity has implications for understanding people, because it suggests an "intrinsic relatedness between the seen bodily acts of others and the internal states of oneself, so that the sensing and representations of one's own movements" (Meltzoff & Moore 1995, p. 54). This primitive self-representation of the body may be the earliest progenitor of the ability to take perspective on oneself as an object of thought. Data indicates that imitation requires a body schema sufficiently developed at the birth to account for the ability to move one's body in appropriate ways in response to environment and especially interpersonal stimuli. The sensory-motor system is, therefore, already predisposed to be coordinated in an orbicular, dyadic communication within which a reciprocal, bidirectional exchange of social information takes place.

There have been many descriptions as to how *early imitation* disappears around the third month of life, giving way to a more mature form of comprehension which consists in the ability to understand the meaning of what is imitated. This secondary form of "intersubjective resonance" includes mechanisms of *shared attention* as well as the *allocentric participation* of movements oriented towards the object (Trevarthen & Hubley 1978). This type of secondary intersubjectivity begins between the ages of three and nine months, with the cooperative use of objects and imitative learning. According to Daniel Stern (2004), learning through allocentric participation allows for *imitation, empathy* and *identification*. These steps correspond to stratified levels of intersubjective attuning that are independent of sophisticated cognitive faculties such as language. They are meant to support superior levels of consciousness. With regards to this, Buck (1994) stresses the spontaneous and biologically determined nature of this form of pre-linguistic communication, defining it an "emotional pre-syntonization" or, in other words, a sort of "conversation between limbic systems". The studies of Krolak-Salmon (2003), revealed that the antecedent region of the insula becomes active when facial expressions of disgust are observed: it is thus plausible that a neural mechanism shared amongst the observer and the observed also exists for emotions, and that this leads to direct experiential comprehension. The empathetic and sensory-motor mechanisms provide the ability to read and feel what others feel and simultaneously allow one to distinguish one's own actions from those belonging to others. According to Antonio Damasio's (1999) interpretation, the understanding of primary emotions is, in fact, correlated to the neural mapping of physical states. Damasio's *somatic-marker hypothesis* proposes a mechanism by which emotional processes can guide behavior, particularly decision-making: both self-perception of one emotions as well as emotional acknowledgement in others depend on two particular structures, the insula and the somatosensory cortex. Therefore, the theory seems to be suggesting that all of our perceptions and actions are accompanied by an immediate sense of possession, that is to say, experienced from a first person perspective.

Developing a sophisticated understanding of others depends on building the capacity for the embodied practices of mind that come to be manifested much earlier than the onset of theory of mind capabilities. These cognitive structures, which contribute to the generation of a primary self-awareness are mature at birth. Thus in the case of neonate imitation, the imitating subject depends on a complex background of embodied processes, a body schema system involving visual, proprioceptive, vestibular and visceral information. Thanks to an empathetic and sensory-modal supramodal system, we can instantly and pre-reflectively gather not only the facial expressions and gestures of others, but they're intentional actions as well, thus experimenting a primary form of *self-knowledge*. The simulation theory as well as the abundance of empirical research regarding the *mirror* areas (Gallese 2003; Rizzolatti & Sinigaglia 2006) support the idea that before meta-representation skills develop, thanks to our perceptive and motor abilities, we are involved in an immediate relationship with others and understand associated mental states and actions.

3 Conclusion

The concept of a *proprioceptive bodily-self* suggested here is quite consistent with what Neisser (1988; 1993) calls an *ecological self* and a primitive self-awareness that is based on both visual proprioception and a sense of movement and action. This fact is not unimportant for the related issues of a sense of self and the perceptions of others. With the notion of an innate intermodal system of body schema, proprioceptive and qualitative experience, it is possible to propose a solution to a problem that many philosophers of mind have attempted to answer: the problem of *what is it like* to be ourselves and the problem of how we know others. So, in this paper we propose to interpret processes of sensory-motor integration in light of the phenomenological approach that allows the definition of pre-reflective self-consciousness. From an evolutionary point of view, we assume that phenomenal consciousness is associated with our perceptive non conceptual experiences and share some properties with our motor and perceptive system.

We suggest that focusing on these aspects of bodily experience may provide a more fruitful framework for understanding the phenomenological proprieties of consciousness, invoking the concepts of a body schema and body image. Specifically, the body schema could involve a pre-reflective experience, while the body image could involve observational consciousness of the body. Experiential and empirical perspectives converge on the idea that pre-reflective bodily self-consciousness corresponds to being functionally and experientially bodily-in-the-world. Both ontogenetic and philogenetic studies suggest the motor-related neuronal processes and structures are integrally linked to sensory and emotive processes, and that much of this integration is organized by motor representations of the body (Panksepp 1998). Thus, simulation literature suggests that people routinely track the mental states of others in their immediate environment. This tracking is apparently done by representing the other's actions in a functionally equivalent way as one's own actions, just a simulation theory predicts. The recent discovery of mirror neurons in the premotor cortex shows a direct link between the motor and sensory system and has important implications for explaining how we understand other people.

So, this development of perceptual and cognitive abilities is enhanced in correlation to a greater amount of crawling and mobility in infancy and more specific perceptual strategies develop when infants are able to execute certain motor abilities (Bushnell & Boudreau 1993). Finally, we assume the hypothesis that the *proprioceptive system* and the *qualitative properties of experience* contributes to the self-organizing development of a neural structure responsible not only for motor action, but for the way we come to be conscious of ourselves, to communicate with others and to live in the surrounding world.

References

Berlucchi, G.; and Aglioti S. (1997): "The Body in the Brain: Neural Bases of Corporeal Awareness". *Trends in Neurosciences*, 20, pp. 560-564.

Bermúdez, J.L.; Marcel, A.; and Eilan, N. (1995): *The Body and the Self.* Cambridge, MA, MIT Press.

Bermúdez, J.L. (1998): *The Paradox of Self-Consciousness.* Cambridge, MA, MIT Press.

Blakemore, S.J.; and Frith, C. (2003): "Self-Awareness and Action". *Current Opinion in Neurobiology, 2, pp. 219-224.*

Braten, S. (1998): *Intersubjective Communication and Emotion in Early Ontogeny.* Cambridge, Cambridge University Press.

Buck, R. (1994): "Social and Emotional Functions in Facial Expression and Communication: The Readout Hypothesis". *Biological Psychology*, 38, pp. 95-115.

Bushnell, E.; and Boudreau, P. (1993): "Motor Development and the Mind: The Potential Role of Motor Abilities as a Determinant of Aspects of Perceptual Development". *Child Development*, 64, pp. 1005-1021.

Chaminade, T.; and Decety, J. (2003): "When the Self Represents the Other: A New Cognitive Neuroscience View on Psychological Identification". *Consciousness and Cognition*, 12, 577-596.

Damasio, A. (1999): *The Feeling of What Happens*. London, Heinemenn (*Emozione e Coscienza*, It. trans. F. Macaluso, Milano, Adelphi, 2000).

Davies, M.; Stone, T. (1995): *Folk Psychology*. Oxford, Blackwell Publishers.

Gallagher, S.; and Cole, D.J. (1995): "Body Schema and Body Image in a Deafferented Subject". *Journal of Mind and Behavior*, 16, pp. 369-390.

Gallagher, S.; and Meltzoff, A.N. (1996): "The Earliest Sense of Self and Others: Merleau-Ponty and Recent Developmental Studies". *Philosophical Psychology*, 9, pp. 213-236.

Gallagher, S. (2001): "The Practice of Mind: Theory, Simulation, Or Primary Interaction?". *Journal of Consciousness Studies*, 5, pp 83-108.

Gallagher, S.; and Jeannerod, M. (2002): "From Action to Interaction". *Journal of Consciousness Studies*, 9, pp. 3-26.

Gallagher, S. (2005): *How the Body Shapes the Mind*. New York, Oxford University Press.

Gallagher, S.; and Zahavi, D. (2008): *The Phenomenological Mind*. Oxford, Routladge (*La Mente Fenomenologica*, It. trans. P. Pedrini, Milano, Cortina, 2009).

Gallese, V.; and Goldman, A. (1998): "Mirror Neurons and the Simulation Theory of Mind-Reading". *Trends in Cognitive Sciences*, 2, pp. 493-501.

Gallese, V. (2003): "The Manifold Nature of Interpersonal Relations: The Quest for a Common Mechanism". *Philosophical Transaction of the Royal Society of London Series B: Biological Sciences*, 358, pp. 517-528.

Gallese, V. (2005): "Embodied Simulation: From Neurons to Phenomenal Experience". *Phenomenology and Cognitive Sciences*, 4, pp. 23-48.

Gallese, V.; and Sinigaglia, C. (2010): "The Bodily Self as Power for Action". *Neuropsychologia*, 48, pp. 746-755.

Goldman, A.I. (1989): "Interpretation Psychologized". *Mind & Language*, 4, pp. 161-185.

Goldman, A.I. (1995): "In Defense of Simulation Theory", in M. Davis and T. Stone (eds.), *Folk Psychology*. Oxford, Blackwell Publishers, pp. 74-99.

Goldman, A.I. (2006): *Simulating Minds. The Philosophy, Psychology, and Neuroscience of Mindreading.* New York, Oxford University Press.

Gopnik, A.; and Meltzoff, A.N. (1997): *Words, Thoughts and Theories.* Cambridge, MA: MIT Press.

Gordon, R.M. (1995): "Developing Commonsense Psychology: Experimental Data and Philosophical Data". *APA Eastern Division Symposium on Children's Theory of Mind*, 27 (www.umsl.edu/~philo/Mind_Seminar/New%20Pages/papers/Gordon/)

Haggard, P. (2005): "Conscious Intention and Motor Cognition". *Trends in Cognitive Sciences*, 6, pp. 290-295.

Harris, P. (1995): "From Simulation to Folk Psychology: The Case for Development", in M. Davies and T. Stone (eds.), *Folk Psychology*. Oxford, Blackwell, pp. 207-231.

Head, H.; and Holmes, G. (1911): "Sensory Disturbances from Cerebral Lesions". *Brain*, 34, pp. 102-254.

Holmes, P.N.; and Spance, C. (2004): "The Body Schema and the Multisensory Representation of Peripersonal Space". *Cognitive Processing*, 5, 94-105.

Jeannerod, M. and Pacherie, E. (2004): "Agency, Simulation and Self-Identification". *Mind and Language*, 2, pp. 113-146.

Krolak-Salmon, P.; and Hénaff, M. (2003): "An Attention modulated response to disgust in human ventral anterior insula". *Annals Of Neurology*, 53, pp. 446-453.

Kugiumutzakis, G. (1998): "Neonatal Imitation in the Intersubjective Companion Space", in S. Braten (ed.), *Intersubjective Communication and Emotion in Early Ontogeny*. Cambridge University Press, pp. 63-88.

Legrand, D. (2006): "The bodily self: The Sensory-Motor Roots of Pre-Reflexive Self-Consciousness". *Phenomenology and the Cognitive Sciences*, 5, pp. 89-118.

Lipps, T. (1903): *Ästhetik. Psychologie des Schönen und der Kunst*. Hamburg, Voss.

Maravita, A.; and Iriki, A. (2004): "Tools for the Body (Schema)". *Trends in Cognitive Sciences*, 8, pp. 79-86.

Meini, C. (2007): *Psicologi per natura: Introduzione ai meccanismi cognitivi della psicologia ingenua*. Roma, Carocci.

Meltzoff, N.A.; and Moore, M.K. (1995): "Infants' Understanding of People and Things", in J.L. Bermùdez (ed.), *The Body and the Self*. Cambridge, MA, MIT Press, pp. 43-69.

Meltzoff, N.A.; and Moore, M.K. (1998): "Infant Intersubjectivity. Broadening the Dialogue to Include Imitation, Identity and Intention", in S. Braten (ed.), *Intersubjective Communication and Emotion in Early Ontogeny*. Cambridge University Press, pp. 47-72.

Meltzoff, N.A. (2002): "Elements of a Developmental Theory of Imitation", in N.A. Meltzoff and W. Prinz (eds.), *The Imitative Mind: Development, Evolution and Brain Bases*. Cambridge University Press, pp. 19-40.

Melzack, R.; Isreal, R.; Lacroix, R.; and Schultz, G. (1997): "Phantom Limbs in People With Congenital Limb Deficiency or Amputation in Early Childhood". *Brain*, 120, pp. 1603-1620.

Metzinger, T. (2003): *Being No One*, Cambridge, MA, MIT Press.

Neisser, U. (1988): "Five Kinds of Self-Knowledge". *Philosophical Psychology*, 1, pp. 35-59.

Neisser, U. (1993): *The Perceived Self: Ecological and Interpersonal Sources of Self-Knowledge*. Cambridge University Press.

Racine, P.T.; and Carpendale, J. (2007): "Shared Practices: Understanding, Language and Joint Attention". *British Journal of Developmental Psychology*, 25, pp. 45-54.

Rizzolatti, G.; and Arbib, M.A. (1998): "Language Within our Grasp". *Trends in Neurosciences*, 21, pp. 188-194.

Rizzolatti, G.; and Sinigaglia, C. (2007): "Mirror Neurons and Motor Intentionality". *Functional Neurology*, 22, pp. 205-210.

Stern, D. (1985): *The Interpersonal World of the Infant: View from Psychoanalysis and Developmental Psychology*. New York, Basic Books.

Stich, S.; and Nichols, S. (1992): "Folk Psychology: Simulation or Tacit Theory?". *Mind and Language*, 7, pp. 35-71.

Trevarthen, C.; and Hubley, P. (1978): "Secondary Intersubjectivity: Confidence, Confiders and Acts of Meaning in the First Year", in A. Lock (ed.), *Action, Gesture, and Symbol*. New York, Academy Press, pp. 183-229.

Trevarthen, C. (1979): "Communication and Cooperation in Early Infancy: A Description of Primary Intersubjectivity", in M. Bullowa (ed.), *Before Speech*. Cambridge University Press, pp. 321-348.

Tsakiris, M.; and Haggard, P. (2005): "The Rubber Hand Illusion Revisited: Visualtactile Integration and Self-Attribution". *Journal of Experimental Psychology: Human Perception and Performances*, 31, pp. 80-91.

Fourth Section
Philosophy of Physics

Physics and metaphysics

Vincenzo Fano
University of Urbino
vincenzo.fano@uniurb.it

The relation between physics and metaphysics is still the object of lively debate (see DiSalle, 2006, pp. 57-58, Ladyman and Ross, 2007, Calosi, 2010 and Dorato, 2010). This is also the consequence of the diffusion in contemporary philosophy of a metaphysical way of thinking, which either does not consider the results of empirical sciences, or else it uses them in a partial and distorted manner (Lowe 2002 and 2006, Sider, 2001 and Varzi, 2001). The notion that metaphysics comes before empirical sciences is an ancient view, given that it already appears in Aristotle (*Met.* E, 1026a, 10ss.), who states that there is a *proto-episteme* (first science) which deals with what is motionless and separate. Indeed Aristotle is careful to underline that the *proto-episteme* neither coordinates nor contains within it all the other disciplines[1]. It follows that his "physics" – in our terms physics, biology and psychology – could not be deduced from the afore-mentioned first science. In the Cartesian perspective, however, physics is the trunk of the tree of philosophy, whose roots are metaphysics (Descartes, *Principes,* 1647, *AT*, IX - 2, 14). Hence, in a certain sense, physics must be derived from metaphysics. The contemporary viewpoint is different. Scholars maintain that metaphysics is a conceptual (a priori) activity independent of physics. Nevertheless this does not mean the latter is

---

[1] Here I follow Alexander of Aphrodisias's interpretation, 447, 30ss. Of *Met.* 1025b 23ss.

derivable from the former. Ladyman and Ross (2007) and Dorato (2010) are against the contemporary perspective.

Before going on we will attempt the following definition:

An assertive sentence is *metaphysical* when it contains predicates which do not appear in the languages of scientific theories.

This definition needs some clarifications. In spite of Putnam's (1962) critiques, the so called "received view" is still the best way to describe scientific theories in their relation with experimental data. According to this perspective, few predicates, which appear in scientific theories, could be defined in operative terms, for instance "length"; others, instead, such as "magnetic field", have a partial empirical meaning, due to the fact that they are part of a theoretical net incompletely linked to experimental data. In fact the link between theoretical terms and experience is only indirect.

I say "almost" operative because we know that no empirical term is completely empirical, since, in order to determine any scientific term, Carnap's (1952) meaning postulates are always necessary. Metaphysical notions are, on the contrary, out of this net.

To sum up, there are terms like "length" that are quite simply connected with the experimental level; others, on the contrary, like "magnetic field", though they are part of the scientific net, are "theoretical terms", that is they acquire empirical meaning only through their link with operative (observative) terms; finally terms like "property" do not belong to any scientific theory. Therefore a sentence in which the latter predicate appears is metaphysical.

Moreover it is necessary to emphasize that my definition is merely negative, that is it demarks metaphysics merely as non-science, but it provides no positive peculiarity of metaphysical sentences. Indeed in this generic sense even sentences like "God is a woman" and "2+2=4" are metaphysical. Furthermore terms like "particle", "action at distance", "genetic information", "biologically altruistic behavior" etc., though they are exactly definable on the basis of a scientific theory, do not belong to any scientific theory. Therefore these kinds of terms in a certain sense are metaphysical, but, being strictly linked to a precise scientific dominion, are on the border between science and metaphysics.

In this definition I partly follow Ladyman and Ross (2007, p. 33), since I determine the boundary between science and metaphysics on the basis of usage and not of either semantic or syntactical peculiarities, as attempted by neopositivist philosophers.

Metaphysics defined in these terms plays an obvious role in science and vice versa. For in the dominion of scientific discovery, as "new philosophers of science" like Popper, Kuhn and Feyerabend showed, metaphysics influences the actual work of the scientist. On the other hand metaphysical speculation is often stimulated by coeval scientific theories, as happened in the cases of Kant and Newtonian mechanics (see Friedman, 1992) and in that of logical positivism and relativistic theories (see Friedman, 1983, chap. 1). But in the present introduction I do not intend to pose the problem of the relation between science and metaphysics in the context of *discovery*, but only in the context of *justification*. It would be reasonable to say that scientific sentences do not need metaphysical justification, but I will not discuss this issue here. On the contrary, my topic is the justification of metaphysical sentences in relation to empirical sciences.

Another point needs clarification: if we suspend our belief in the existence of something external to space-time and in general in entities not described and explained by our best scientific theories, we can say that from the outset metaphysics does not possess a field of objects peculiar to it. This does not mean that nothing of the kind exists – possible worlds, universals, souls etc. – but that such possible existence could not be assumed before scientific investigation. That is, scientific results could persuade us that there are, as it were, such extra-scientific entities. But, again, establishing their peculiarities, though a metaphysical task, must be based on outcomes of empirical sciences. If I am right, statements like "properties are characteristics of an object" are not allowed, but only "we define the term 'property' along the following lines:…". Thereafter we establish whether it is possible that these kinds of metaphysical entities exist on the basis of the results of empirical sciences. In other words, in metaphysics concepts could not be described, but must be defined. The definition of metaphysical terms is a really finely tuned task, which, though in the phase of discovery certainly uses scientific results[2], in the phase of justification must be accomplished completely a priori.

Beside these first considerations, it seems opportune to outline the relation between science and metaphysics preserving the following four issues:

1. Doing metaphysics is a different activity with respect to doing science. This is strongly suggested by the fact that almost no scientist would accept that his work being dubbed "metaphysical" in the previously outlined

---

[2] Think again of the case of Kant and Newtonian physics. The significance of this kind of activity is overlooked by Ladyman and Ross, 2007. It should not be confused with what Strawson (1959) calls "descriptive metaphysics", that is the analysis of the subjective structure of knowledge. Metaphysics in such a sense is a useful and correct work; but here we are discussing the recent proposal of what Strawson would have called a "revisionary" metaphysics.

sense. Vice versa, though metaphysicians take part in the global cognitive enterprise we can call science-and-philosophy, one could not state that they either spend their time in a laboratory or formulate theoretical models.

2. Except for logical and analytical sentences[3], it is not possible to establish a priori the truth of a metaphysical sentence, albeit in a partial and probable way. That is to say, empirical sciences, and only they, could provide arguments supporting any metaphysical thesis. By the locution "empirical sciences" I mean all sciences, not only physics, as is sometimes intended. Physics concerns the smallest and simplest objects, but, even if what is bigger and more complex were completely determined by what is smallest and simplest, contemporary physics would not be able to accomplish such a reduction. Some people believe that such an impossibility is due only to the complexity of calculation, but maybe – and the history of science supports this hypothesis – other important conceptual revolutions lie in the future of our investigation. Therefore all sciences aiming at the discovery of natural laws can provide arguments supporting a metaphysical thesis, from physics to sociology, through chemistry, biology and psychology[4]. This latter thesis derives from Kant, according to which there is no kind of intellectual intuition. That is, our knowledge proceeds only by formulating models, and not by any form of direct access to some not empirical ideality. Here is an example. Some people maintain that in the realm of metaphysics one can establish a priori *all* possible kinds of entity (Lowe, 2006). While the possibility to outline a priori different kinds of entity, which could be useful for the metaphysical investigation of scientific results, is reasonable, nevertheless in a purely metaphysical context it is not possible to state the truth of the sentence "*x,y,z...* are all possible kinds of entity". This is so for a simple reason, which becomes evident through an easy example: who could have imagined an object like Minkowskian space-time before Einstein? But today we have grounds to believe that far from gravitational fields and for bodies moving with velocity comparable to that of light, space-time has precisely a Minkowskian structure. In other words, in their continuous evolution, empirical sciences propose ever-new kinds of objects, which cannot be foreshadowed a priori by metaphysics.

---

[3] Here the discussion as to whether analytic sentences actually exist is not relevant.

[4] In any case, the principle stated by Ladyman and Ross (2007, 44) "*Primacy of Physical Constraint*" seems sensible: if a law L belonging to a science different from physics is against contemporary physical laws, then L must be refused. Nevertheless we cannot consider this constraint as absolute; consider for instance Kelvin's criticism of Darwin's theory based on an erroneous estimation of the sun's age.

3. Even though endowed with empirical meaning, metaphysical sentences are not falsifiable and therefore neither confirmable (Watkins, 1958, Tarozzi, 1988). That is, the relation between scientific theories and metaphysical sentences does not have a logical character (see Dorato, 2010, p. 6). Therefore the results of scientific research could not provide a definitive falsification of metaphysical theses[5] – as for instance was presumed in the celebrated case of the experimental violation of Bell's inequality[6], which would have falsified Einstein's local realism – but in the best of cases they might produce good arguments against a metaphysical issue. There is another argument favoring my statement.

The "New philosophy of science" since Kuhn has denied the possibility of finding experimental results which can definitely falsify an abstract and general scientific theory. Nevertheless, with suitable caveats, it is probably possible to outline a cognitive rationality holding in the evaluation of scientific theories, at least from a normative point of view. In spite of this, it is well known that in practice scientists seldom respect the methodological rules which they themselves maintain in principle; and sometimes it is even better so. In any case the concept of falsification has not completely lost its epistemological relevance in science. But, if it is so difficult to find a clear falsification procedure for fundamental scientific theories, it seems utopian to do the same for metaphysical hypotheses.

4. Often a distinction is drawn between "good" and "bad" metaphysics; some people, for instance Carnap, maintain that all metaphysics is bad. It seems to me more interesting to emphasize that there are metaphysical questions which are more or less interesting for a given group of persons. Sprenger, Institor and their contemporaries, for instance, considered the question of whether a woman is or is not possessed by a devil so important that they wrote a handbook on the topic, *Malleus maleficarum* (1486). Almost always and for almost all people it has been very interesting to know whether there is life after death or not. And it would be easy to find many other similar examples. But the very point is that available scientific knowledge allows a distinction between metaphysical questions to which it is possible to give an answer, even though provisional and revisable, and metaphysical questions, which are so far from scientific answers that till now and maybe forever we are not able to debate them even in a partially

---

[5] It would be necessary to establish whether a predicate in the course of research could migrate from a metaphysical to a scientific status. If so, empirical science could definitely falsify metaphysical theses, but only if these theses have been transformed into scientific sentences.

[6] See, for instance, Cohen, Home and Stachel (1997), a collective volume significantly entitled "*Experimental metaphysics*".

justified way[7]. For instance the discovery of so many scientific laws in most different domains of objects is a strong ground favoring the existence of some kind of instantiated universals, even though it is difficult to establish in mere metaphysical terms the nature of such universals, as many have attempted by means of rough terms like "property", "disposition" and "causal power". On the contrary contemporary science gives no important suggestion regarding the possible existence of an infinitely good God.

Before concluding I would like to introduce three methodological principles which seem to me essential for performing metaphysics on the basis of scientific results.

I. Metaphysical conclusions, which we derive from scientific investigation, have a low *weight* and are revisable. The concept of "weight" was outlined by Keynes (1921) and it is easily explained through the following example. Moving from a partial number of counted ballot-papers, we have to establish the percentage of electors who have chosen the center-left party and that of those who have chosen the center-right party. Two hours after the closure of the polls, it is possible to estimate through a suitable algorithm, that the center-left receives 45% of votes, whereas the center-right 37%, and only 13% of votes have been counted. After six more hours, on the contrary, 67% of the votes have been counted and the same algorithm predicts that the center-left has decreased to 41% and the center-right increased to 41%. We can say that the first probabilistic evaluation – based on only 13% of counted votes – has less *weight* than the second one, which, instead, is based on 67% of counted votes. Therefore, since metaphysical conclusions are very general and abstract, it is evident that they have low weighted justifications. Moreover, as happens with every thesis maintained on the basis of arguments, metaphysical conclusions are revisable.

II. Given a domain of objects D, we often have what we might dub "the best explanation of D". In order to do metaphysics from scientific results, we have to assume a sort of "partial inference to the best explanation", i.e. we have to allow the validity of arguments of the following kind: if S is the best explanation of D, then S is at least partially true of D. If moreover we assume a correspondence theory of truth, we can say that S catches at least

---

[7] The *Principle of Naturalistic Closure* proposed by Ladyman and Ross (2007, 37), deserves some further discussion. It affirms that a metaphysical statement could be reasonably discussed if and only if its possible truth would increase the explicative power of at least two specific scientific hypotheses, one of which must come from physics. At first sight, this principle seems a sufficient condition but not a necessary one.

partially how D is constituted. I know that it is very difficult to establish exactly what the term "partially" means in this context, but however this kind of inference is reasonable. I further underline, incidentally, that regarding the *vexata quaestio* of what is truth for scientific theories, it is possible to support with good arguments that when searching for truth "coherentism" is unavoidable, since our investigation is always based on a comparison between sentences, but the final condition of truth is the correspondence of the sentences with reality.

III. The first two meta-philosophical principles I have proposed are well-known, even though partially controversial; the next one is, instead, less common. It is not possible to deny completely the cognitive value of sensations, otherwise, as Democritus had already observed (*DK*, B 125), despite being a staunch rationalist, the whole building of science would collapse, since, though built on piles (Popper), it lies *also* on perception. Nevertheless the best scientific theories often propose a scientific image different and sometimes opposed to the manifest one (Sellars). Therefore the following question arises: how can one evaluate this conflict? Who is right between common sense and science? I believe that to answer such a query we have to use the following criterion: if we have a good scientific explanation of why we perceive the world in so different a way with respect to what scientific theories suggest, then we have to credit the latter. Otherwise we have to credit perception, till there is no contrary evidence. If this kind of "moderate scientism" holds, as we might dub it (see Angelucci and Fano, 2009), then the following principle holds as well: we have to accept the existence only of the theoretical entities introduced by science, where we know either why we don't perceive them or why we perceive them differently with respect to how the best scientific models describe them.

To conclude I would like to recall the 4.111. proposition of Wittgenstein's *Tractatus*, where the Austrian philosopher affirms that philosophy is not a natural science, i.e. it does not lie side by side with them, but is placed either before or after them. For Descartes metaphysics came *before* the sciences; for many contemporary philosophers, metaphysics is *independent* of science; in the conception I have here briefly outlined, metaphysics instead comes *after* the sciences and assembles them all together in an image as coherent as possible.

I thank Mario Alai, Claudio Calosi and Giovanni Macchia for their suggestions given after reading a first draft of the paper.

References

Angelucci, A.; Fano, V. (2009): "On What There Really Is. Empirical Realism Between Physics and Psychology". *Teorie e modelli*, 13, pp. 105-118.

Calosi, C. (2010): "Introduction. Physics and Metaphysics". *HumanaMente*, 13, pp. III-XVII.

Carnap, R. (1952): "Meaning Postulates". *Philosophical Studies*, 3, pp. 65-73.

Cohen, R.S.; Home, M.; Stachel, J.J. (1997): *Experimental Metaphysics*. Berlin, Springer.

DiSalle, R. (2006): *Understanding Spacetime. The Philosophical Development of Physics from Newton to Einstein*. Cambridge, Cambridge University Press.

Dorato, M. (2010): "Physics and Metaphysics. Interaction or Autonomy?". *HumanaMente*, 13, pp. 1-11.

Friedman, M. (1983), *Foundation of Space-Time Theories*. Princeton, Princeton University Press.

Friedman, M. (1992): *Kant and the Exact Sciences*. Cambridge Mass., Harvard University Press.

Keynes, J.M. (1921): *A Treatise on Probability*. London, McMillan.

Ladyman, J.; Ross, D. (2007): *Everything Must Go. Metaphysics Naturalized*. Oxford, Oxford University Press.

Lowe, E.J. (2002): *A Survey of Metaphysics*. Oxford, Oxford University Press.

Lowe, E.J. (2006): *The Four-Categories Ontology. A Metaphysical Foundation for Natural Science*, Oxford, Oxford University Press.

Putnam, H. (1962): "What Theories Are Not", in E. Nagel, P. Suppes and A. Tarski, ed., *Logic, Methodology and Philosophy of Science*. Stanford, Stanford University Press, pp. 215-227.

Sider, T. (2001): *Four-Dimensionalism. An Ontology of Persistence and Time*. Oxford, Oxford University Press.

Strawson, P. (1959): *Individuals. An Essay in Descriptive Metaphysics*. London, Methuen.

Tarozzi, G. (1988): "Science, Metaphysics and Meaningful Philosophical Principles". *Epistemologia,* 11, pp. 97-104.

Varzi, A. (2001): *Parole, oggetti, eventi e altri argomenti di metafisica*. Roma, Carocci.

Watkins, J. (1948): "Confirmable and Influential Metaphysics". *Mind*, pp. 344-365.

On spacetime coincidences[*]

Giovanni Macchia
University of Urbino
lucbian@hotmail.com

He is not a true man of science who does not bring some sympathy to his studies, and expect to learn something by behavior as well as by application. It is childish to rest in the discovery of mere coincidences, or of partial and extraneous laws. The study of geometry is a petty and idle exercise of the mind if it is applied to no larger system than the starry one.
Henry David Thoreau[**]

In this work I just wish to clarify an *apparent* contrast on the nature of two possible descriptions of spacetime. They belong to two contexts: one derives from a general argument – the so-called *Point-Coincidence Argument* (PCA) – set up by Einstein; the other from the foundations of cosmology as established by the so-called *Weyl's Principle* or *Postulate* (WP). At the centre of both approaches there is the notion of *spatiotemporal coincidence*, or *coincidence of worldlines*. But with a completely different

[*] Many thanks to Claudio Calosi, Vincenzo Fano and Pierluigi Graziani for having read and commented this paper. It goes without saying that the conclusions advanced here are entirely my own.
[**] *A Week on the Concord and Merrimack Rivers* (1849), in *The Writings of Henry David Thoreau* (Vol. 1, p. 387), Houghton Mifflin Company, Boston, 1906.

perspective. In the first approach, the Einsteinian one, this notion is fundamental because it is precisely the full realization of these "meetings" that *relationally* constitutes spacetime. In the second approach, the Weylian one, these coincidences *must* be completely absent in order for a *relational* set-up of spacetime to make sense.

This paper is divided into three main sections, the first two completely separate and independent from one another. Section 1 introduces Einstein's PCA and analyzes the peculiarities of its ontological claims about spacetime. Section 2 outlines WP, its role in the foundations of models of standard cosmology, and its possible repercussions on the ontology of cosmic spacetime. Finally, section 3, after having highlighted some inconsistencies in the PCA, tries to explain why the preceding two sections have been closed with such different ontological conclusions.

1 Foundations of general relativistic spacetimes

1.1. The *hole argument*

A very important and tiring step towards the completion of General Relativity has been the famous *hole argument*. Such an argument led Einstein at first to reject generally covariant field equations for his gravitational theory, and then led him to reinstate general covariance thanks to another argument he himself developed: the PCA. This is a story by now told many times, in particular since Earman and Norton, in 1987, revived Einstein's original hole argument with a new modern lymph.[1] So here I will very briefly depict, in order to arrive at the PCA, only the essential traits of Einstein's original hole argument, which he presented first in the late fall of 1913.

In this argument Einstein analyses the gravitational field within an open region of spacetime (the hole) devoid of matter (the stress-energy tensor $T_{\mu\nu}$ vanishes). $g_{\mu\nu}$ represents the metric tensor field (satisfying the field equations) in a coordinate system x_τ. What happens physically in the hole is then completely determined by $g_{\mu\nu}$. He then introduces a new coordinate system x'_τ which agrees with the coordinate system x_τ *only* outside the hole but comes smoothly to differ from it within the hole. $g_{\mu\nu}$ can be expressed in terms of this new coordinate system: the transformed expression, obtained by the usual tensor transformation law, is $g'_{\mu\nu}$.

---

[1] See Earman and Norton (1987).

Therefore, $g_{\mu\nu}(x_\tau)$ and $g'_{\mu\nu}(x'_\tau)$ represent the same gravitational field: both can be considered as symmetric matrixes given by a set of ten functions of the variables x_τ and x'_τ respectively.

It is now possible to construct a new set given by the functions of the "new" matrix $g'_{\mu\nu}$ considered as functions of the "old" coordinates x_τ. In such a way one obtains $g'_{\mu\nu}(x_\tau)$. If the field equations for the metric tensor are *generally covariant*, namely they are form-invariant under general coordinate transformations[2], then – Einstein realizes – they must be satisfied also by $g'_{\mu\nu}(x_\tau)$. But, the problematic point – Einstein concludes – is that $g_{\mu\nu}(x_\tau)$ and $g'_{\mu\nu}(x_\tau)$ are different from each other ($g_{\mu\nu}$ and $g'_{\mu\nu}$ are different functions), i.e. they represent *distinct* gravitational fields in the same coordinate system, in spite of the fact that outside and at the boundary of the hole they coincide.[3] So the hole argument shows that general covariant field equations allow as solutions the distinct gravitational fields g and g'. In other words, no specification of the metric field outside of and on the boundary of the hole could *uniquely* determine, by generally covariant differential equations, the field inside the hole, in such a way undermining determinism (that Einstein called "law of causality").[4]

The aim of the PCA is exactly that to give a grounds to accept that $g_{\mu\nu}(x_\tau)$ and $g'_{\mu\nu}(x_\tau)$ represent the *same* gravitational field.

1.2. The point-coincidence argument

Einstein realized, before the end of 1915 when he completed his theory of General Relativity, that the crucial error he made in the hole argument was an illicit assumption regarding the individuation of the manifold points.

---

[2] In general, a metric $g_{\mu\nu}(x)$ is form-invariant under a transformation from x_τ to x'_τ if $g'_{\mu\nu}(x')$ is the same function of x'_τ as $g_{\mu\nu}(x)$ is of x_τ.

[3] Within the hole, numerically equivalent values of x_τ and x'_τ label *supposedly* different points of the manifold. This implies that numerical value of $g'_{\mu\nu}$, initially assigned to the point x'_τ, is now assigned to the point x_τ, so that, in general, $g'_{\mu\nu}(x_\tau)$ will differ from the numerical value of $g_{\mu\nu}$ originally assigned to the point x_τ. For this reason Einstein concludes that, in general, $g_{\mu\nu}(x_\tau)$ and $g'_{\mu\nu}(x_\tau)$ differ.

[4] Note that Einstein saw a problem for the law of causality in the inequality $g_{\mu\nu}(x_\tau) \neq g'_{\mu\nu}(x_\tau)$ and not, as some commentators have maintained, in the $g_{\mu\nu}(x_\tau) \neq g_{\mu\nu}(x'_\tau)$. The latter is, in fact, a trivial coordinate transformation. For detailed historical accounts see Norton (1984) and Stachel (1989).

He had assumed, in fact, that the "two" solutions $g_{\mu\nu}(x_\tau)$ and $g'_{\mu\nu}(x_\tau)$ found in the hole were different precisely because they assigned different values of the metric *to one and the same point*. In other words, he thought that the mere coordinatization of the manifold was sufficient for individuating its points.[5] But it is not: coordinatization does not offer an invariant scheme of individuation which instead is yielded by what will be called, by Stachel in 1980, PCA.

Permit me to quote at length the famous crucial passage by which Einstein officially "closes" his struggle with the general covariance (in the disguise of hole argument), by definitively accepting it in his field equations, and delineates the PCA:

> That this requirement of general covariance, which takes away from space and time the last remnant of physical objectivity, is a natural one, will be seen from the following reflexion. All our spacetime verifications invariably amount to a determination of spacetime coincidences. If, for example, events consisted merely in the motions of material points, then ultimately nothing would be observable but the meetings of two or more of these points. Moreover, the results of our measurings are nothing but verifications of such meetings of the material points of our measuring instruments with other material points, coincidences between the hands of a clock and points on the clock dial, and observed point-events happening at the same place at the same time.
>
> The introduction of a system of reference serves no other purpose than to facilitate the description of the totality of such coincidences. We allot to the universe four spacetime variables x_1, x_2, x_3, x_4 in such a way that for every point-event there is a corresponding system of values of the variables $x_1 \cdots x_4$. To two coincident point-events there corresponds one system of values of the variables $x_1 \cdots x_4$, i.e., coincidence is characterized by the identity of the coordinates. If, in place of the variables $x_1 \cdots x_4$ we introduce functions of them x'_1, x'_2, x'_3, x'_4, as a new system of coordinates, so that the systems of values are made to correspond to one another without ambiguity, the equality of all four coordinates in the new system will also serve as an expression for the spacetime coincidence of the two point-events. As all our physical experiences can be ultimately reduced to such coincidences, there is no immediate reason for preferring certain systems of coordinates to others, that is to say, we arrive at the requirement of general covariance. (Einstein 1916, p. 117-118)[6]

---

[5] As Norton (1993, p. 805) points out: "In executing the hole argument, in order to effect the transition from $g_{\mu\nu}(x_\tau)$ to $g'_{\mu\nu}(x_\tau)$, one has to assume, in effect, that the coordinate system x_τ, has some real existence, independent of the $g_{\mu\nu}$ or $g'_{\mu\nu}$. For, figuratively speaking, one has to remove the field $g_{\mu\nu}$, leaving the bare coordinate system x_τ, and then insert the new field $g'_{\mu\nu}$".

[6] At first sight it seems that Einstein simply equated general covariance with the mere arbitrariness of the choice of coordinates (in modern mathematical terms, such a fact is called *invariance under passive diffeomorphisms*). Actually, as Stachel first revealed in

The lesson of the PCA is that there is no physical reality attaching to a coordinatization: coordinates have become physically meaningless parameters. Whenever we have two different solutions $g_{\mu\nu}(x_\tau)$ and $g'_{\mu\nu}(x_\tau)$, with respect to the same coordinate system, such a fact does not have a physical content, i.e., attributing *different* solutions to the same manifold is senseless insofar as that difference should be based on manifold points endowed with an identity that coordinatization is not able to give them. The physical content of General Relativity is fully exhausted by the catalogue of the spacetime coincidences. If an event is defined by the intersection of the worldlines of two infinitesimal test particles, or two light rays (or also one particle and one light ray), the significance of this intersection is quite independent of any coordinate system. Indeed, two events intersect if they correspond to equal values of the coordinates, and a coordinate transformation changes these values but *not* their equality. Therefore, any transformation that preserves those equalities preserves the physical content. In other words, field equations of General Relativity do not have to determine the metric field and its geodesics uniquely, but only *intersections* of geodesics, i.e., point-coincidences.[7]

1.3. *Point* coincidence or *pointer* coincidence?

An important distinction within the PCA must be pointed out. This can be intuited from the following two excerpts from Einstein's letters. The first one to Ehrenfest (December 26, 1915) and the second one to Besso (January 3, 1916). To Ehrenfest, he writes:

> The physically real in the universe (in contrast to that which is dependent upon the choice of a reference system) consists in *spatiotemporal coincidences*.[8] Real are, e.g., the intersections of two different worldlines, or the statement that they *do not* intersect. [...] All spatiotemporal point coincidences [...], i.e., everything that is observable".

1980, he was thinking of what today are called *active diffeomorphisms*, namely those much more significant functions that map points of manifold to points of manifold. See Norton (1989) for an enlightening clarification of Einstein's use of coordinate systems and covariance principles translated into modern physical-mathematical terms.

[7] See also a *gedankenexperiment* from a letter of Einstein to Ehrenfest of January 1916, where Einstein makes it clear how his PCA works (in Norton 1987, pp. 173-174). For a more modern technical exemplification of PCA see Rovelli (2004, pp. 47-51).

[8] And in a footnote Einstein adds: "And in nothing else!".

To Besso:

> From a physical point of view, nothing is *real* except the totality of spatiotemporal point coincidences. [...] The meeting of the points, i.e., the points of intersection of [...] worldlines, would be the only reality, i.e. observable in principle".[9]

It is evident that in the longer of Einstein's quotations in sect. 1.2, the PCA assumes a more *epistemological* aspect (Einstein seems to stress mostly the question of the *observability*). Instead, in the initial parts of these last two excerpts, PCA also manifests a clear *ontological* content (the invariance of coincidences is what qualifies them as *physical reality*). In this regard, Howard (1999) distinguishes between *pointer coincidences* (those of epistemological flavor) and *point coincidences* (the ontological ones), insofar as the former are the "spatiotemporally extended intersections of the world tubes of 'observable' objects such as the needle on an electrometer and a mark on its scale" (p. 464), whereas the latter are the, in principle, unobservable infinitesimal intersections of possible worldlines.

According to Howard – and it is really hard to not agree with the reasons expounded in his profound essay – PCA, at least in Einstein's understanding, primarily concerns infinitesimal point coincidences, *not* finite pointer coincidences, that is to say, the invariant content of a generally-covariant spacetime theory is wholly determined by the former (that make up the set of all intersections of possible worldlines) and not by the latter. Furthermore, Howard argues that such a role of point coincidences is played by virtue of their invariance under arbitrary transformations, not their observability, insofar as this invariance is a necessary but not a sufficient condition for the relevant kind of observability. In a few words, Howard disentangles the association between invariance and observability that has sometimes led to misinterpretations of PCA, in which ontological questions were blended with epistemological ones. On the other hand, that Einstein himself was more interested in ontological questions than in epistemological ones, is clear from his pronunciations on the observability of spacetime coincidences, which are indeed almost an afterthought – underlines Howard –, an addendum to the primary assertion concerning the physical reality of point coincidences, not involving a positivistic definition of the real via observability. As Howard lastly declares: "A reading closer to Einstein's would be that point coincidences are observable only because they are real, observability being

---

[9] See Howard (1999, p. 468) and Stachel (1989, p. 86) for the complete texts of these letters.

a characteristic feature of the real, but not part of its very definition" (1999, p. 494).

Reichenbach focuses on another aspect that can be regarded, in a certain sense, as a further extension of Howard's reasoning. In a passage from *Philosophie der Raum-Zeit-Lehre*, Reichenbach criticizes those who want to conflate *objective* spatiotemporal point coincidences with *subjective* coincidences in sense experience:

> It is a serious mistake to identify a coincidence, in the sense of a point-event of the space-time order, with a coincidence in the sense of a sense experience. The latter is *subjective coincidence*, in which sense perceptions are blended; for instance, the experience of sound can be blended with the impression of light. The former, on the other hand, is *objective* coincidence, in which physical things, such as atoms, billiard balls or light rays collide and which can take place even when no observer is present. The space-time order deals only with objective coincidences. (1928, p. 286)

It goes without saying that pointer coincidences do *not* coincide (no pun intended!) with sense-experience coincidences. The former, however, being of epistemological flavor, in some cases could actually be "weaker", with respect to the point coincidences, because of their characters of subjectivity.

To sum up the PCA, I still rely on Howard's paper (p. 470), which distinguishes three features deducible from this argument:
1) *nothing* is real except the totality of spatiotemporal point coincidences;
2) such point coincidences are observable in principle;
3) a new invariant scheme of individuation for the points of the spacetime manifold is implicitly yielded (a fact not explicitly stressed by Einstein).

The second feature – on which I have already spent some words – does not pertain to the subject of this paper, so I shall set it aside.[10]

---

[10] Howard (1999) considers it less essential, even if it has grown in importance in the more recent secondary literature. He disputes Friedman's old claim (actually, Friedman has recently retracted it) according to which the PCA "represents the birth of the modern observational/theoretical distinction" and therewith "the beginnings of the empiricist and verificationist interpretations of science characteristic of later positivism" (1983, p. 24). But *only* the beginnings, Friedman specifies: "Not all 'space-time coincidences' are literally observable: consider the collision of two elementary particles. [...] Perhaps what we should say is this: with respect to geometrical structure, the observable = the totality of space-time coincidences" (*ibid.*). Coherently, Howard also comments on and approves of Schlick's anti-empiricist interpretation. In any case, we can more generically say, with Lusanna and Pauri (2003, p. 4), that the PCA "offered mainly a pragmatic solution" to the hole argument's problems and that it "was based on a very idealized model of physical

According to the third feature (a consequence of the first one) two manifold points are physically considered the same if and only if they are constituted by the same spatiotemporal coincidence. It is such a physical individuation that allows us to avoid hole argument traps.

To the first and most important – at least for what I am going to say – feature is dedicated the next subsection.

1.4. PCA implications for the spacetime ontology

As regards the ontology established by PCA there have been somewhat uncharitable judgements. For instance, Earman and Bergia. The former, in partial agreement with Friedman's "first version", accuses PCA of being a "little disappointing in its reliance on a crude verificationism and an impoverished conception of physical reality" (1989, p. 186). The latter underlines that the PCA solution is "epistemologicamente raffinata ma ontologicamente debole. Non può infatti che essere sentita come debole un'ontologia che riduce lo spaziotempo a pura sede di coincidenze spaziotemporali"[11] (1995, p. 61). But it is not on this general subject that I wish to linger. My interest is in the ontology that spacetime inherits from PCA.

It is usually appropriately maintained that the PCA favors a *relational* view[12] insofar as the material points – i.e., those idealized point-unextended objects constituted by the intersections of worldlines – are considered prior to spacetime.

This is the opinion that can be drawn from Einstein's words themselves. When he speaks about the "last remnant of physical objectivity" taken away from space and time (see quote in sect. 1.2), he is indeed referring to the fact that: "L'*oggettività* (qui coincidente con *realtà*) dello spazio e del tempo

---

measurement where all possible observations reduce to the intersections of the worldlines of observers, measuring instruments, and measured physical objects".

[11] "Epistemologically sophisticated but ontologically weak. Indeed, an ontology that reduces spacetime to a pure place of spatiotemporal coincidences can only be perceived as weak".

[12] Here I am not interested in Einstein's "true" philosophical positions, but only in the ones deducible from PCA. On the other hand, at least as regards the ontology of spacetime, the first goal is not an easy subject as, in the course of his life, Einstein expressed opinions wavering between interpretations favoring both relationism and absolutism. However, roughly speaking, a certain shifting from a Machian view towards an absolutist conception can be recognized in his thought.

sarebbe stata – per effetto di tale teoria [General Relativity] – totalmente ricondotta a *relazioni fra i 'corpi'*"[13] (Pauri 1996, p. 103).

This conclusion is better specified by Stachel (1989). He reads PCA in favor of the fact that, according to Einstein, the points of spacetime cannot be physically distinguished except by the properties and relations induced by matter or, where no matter is present, by the metric field: "Only a physical process can individuate the events that make up space-time [...] A manifold only becomes a space-time with a certain gravitational field after the specification of the metric tensor field, and that, prior to such a specification, there is no physical distinction between the elements of the manifold" (1989, pp. 87-88). However, Einstein's comment on the last vestige of physical objectivity was not meant "to indicate that space and time have *no* physical reality, but that they no longer have any *independent* reality, apart from their significance as the spatial and temporal aspects of the metrical field" (*ibid.*).[14]

Lastly, a further quote which gives, in a few precise words, an accurate picture of how PCA quietly flows into a relational sea, above all when matter is present: "Since now space-time coordinates have no objective meaning, *bodies* and light-rays first of all have to define, i.e. to *individuate*, points and instants, by conferring their identity upon them by means of coincidences, thus enabling them to *serve* as the *loci* of other bodies and events" (1991, p. 319; italics original).

1.5. A first conclusion

From PCA it follows that the fundamental ontology of general relativistic spacetimes is given by a spacetime manifold whose points are nothing other than the intersection-points of worldlines of more or less idealized material objects (particles, light rays, measuring apparatus, and so).

---

[13] "The objectivity (here coinciding with reality) of space and time would – as a result of this theory [General Relativity] – have been totally reduced to relations among bodies".

[14] A quick clarification is needed. Relationism here descends from the fact that the metric field is considered "matter-like" – indeed Stachel thinks that a "gravitational field is just as real a physical field as any other" (1993, p. 144) – whereas the manifold is taken as the underlying "space-like" substratum. *But*, if the metric field were considered "space-like" – *à la* Hoefer (1996), for example, who considers manifold points physically existing (even though devoid of primitive identity) independently of the metric – also a *metric field substantivalism* could make sense. Not only, a *structuralist* position has also been proposed (see, for instance, Torretti 1983, p. 167). However, in this context it is impossible to weigh up the pros and cons of the multitude of subtleties that the recent philosophical literature on spacetime has bravely produced.

In short: spacetime is *relationally* constituted by these intersections.

2 Foundations of standard cosmological spacetimes

Two important principles are placed at the foundations of standard cosmology: the *Cosmological Principle* (CP) and *Weyl's Principle* (WP). The former can be found in all cosmology textbooks, the latter is often completely neglected.

Let us take a look at the nature of cosmic spacetime when standard cosmological models are approached by giving a special prominence to WP.

2.1. Some remarks on Weyl's Principle

In Weyl's 1926 formulation[15] – referred to de Sitter's universe, then by him considered as the most satisfying cosmological model – WP states:

> The worldlines of the stars form a sheaf, which rises in a given direction from the infinitely distant past, and spreads out over the hyperboloid in the direction of the future, getting broader and broader. (quoted in Goenner 2001, p. 121)

Weyl is suggesting that the distribution of stars could be described by a congruence of non-intersecting timelike worldlines (a family of non-crossing curves which fills spacetime), diverging (the universe is expanding) from a common point in the past.[16]

Nowadays, clusters of galaxies (or even clusters of clusters!) are taken as the elementary constituents of the expanding universe because it is these giant agglomerates of matter that follow the Hubble expansion pattern quite closely. Although clusters form a discrete set, one can extend it to a *continuum* by a smooth-fluid approximation. The idea is that one averages the speed of matter in a given large-scale region of the universe and assigns that speed and the mass of that region to a fictitious entity called *fundamental particle* (one can imagine it placed at the center of mass of that region). Fundamental particles are *freely falling* insofar as their motions are affected by no forces except gravity and inertia. These material particles, when regarded as mere geometric points, constitute the kinematic

---

[15] For historical aspects see Bergia (1991), Bergia and Mazzoni (1999), Ehlers (2009), Goenner (2001), Kerszberg (1986), North (1965), Rugh and Zinkernagel (2011).

[16] The "*infinitely* distant past", that obviously clashes with the big bang theory, is negligible. On the other hand, Weyl himself specifies that "all beginnings are obscure" (Weyl 1922, p. 10).

substratum of the model. Each point is crossed by only one worldline. Attached to each particle a locally inertial reference can be conceived, so that all the matter of that region is at rest relative to that frame. All these frames form a sort of global *comoving* (i.e. moving with the expanding motion of matter) reference frame.

Thus WP stipulates that the *large scale* motions of fundamental particles are highly streamlined: no randomness in their motions, no vorticities in their trajectories, no collisions among particles (*except* at a singular point, the common "origin", in the past) (see fig. 1a). Furthermore, the bundle of worldlines is thought to have been *causally interconnected* since its origin (Weyl 1930, p. 938).[17] The importance of this common origin and, above all, of such a "regularity" of the worldlines is that they provide *natural synchrony calibration* for all events (the intersection theoretically defines the zero of time).

This guarantees that spacetime be globally resolved into space *and* time, i.e. that it can be foliated in a sequence of "space slices", orthogonal to the bundle, whose succession instantiates the flow of what is called *cosmic time*.[18]

It is important to note, however, that galaxies, planets and all the other *small*-scale objects have their *proper* motions that follow worldlines interweaving like the fibers in a rope. So real *local* motions are, in a sense, chaotic (see fig. 1b). Notwithstanding, they have very low velocities (less than one-thousandth of the velocity of light c), so they are negligible when compared to large-scale velocities of clusters (comparable to c). These low velocities allow us to pass from a smooth-fluid approximation to a dust approximation (the simplest situation with pressure $p \approx 0$) in which dust moves geodesically (Rindler 2006, p. 301).[19]

---

[17] This assumption, according to which all curves share the same causal past and future, makes the universe a coherent whole: WP is "capable of relating all the parts of the universe to one another" (Kerszberg 1989, p. 4). In Ehlers' opinion (1990, p. 29; 2009, p. 1657) this is the first use of causal structure in General Relativity. Note that, in modern mathematical terms, the vorticity-free claim is equivalent to the requirement of *stable causality* for spacetime (see Earman 1995, par. 6.3).

[18] Mathematically speaking, this means that the line element $ds^2 = g_{\mu\nu} dx^\mu dx^\nu$ becomes $ds^2 = c^2 dt^2 + \sigma_{\mu\nu} dx^\mu dx^\nu$. I have spoken of "flow" in a generic neutral sense, without presupposing a real flow of time in the tensed sense of the A-theorists of time.

[19] For matters concerning the early epochs of the universe when dust approximation breaks down see Rugh and Zinkernagel (2009; 2011) and Narlikar (2002, p. 130).

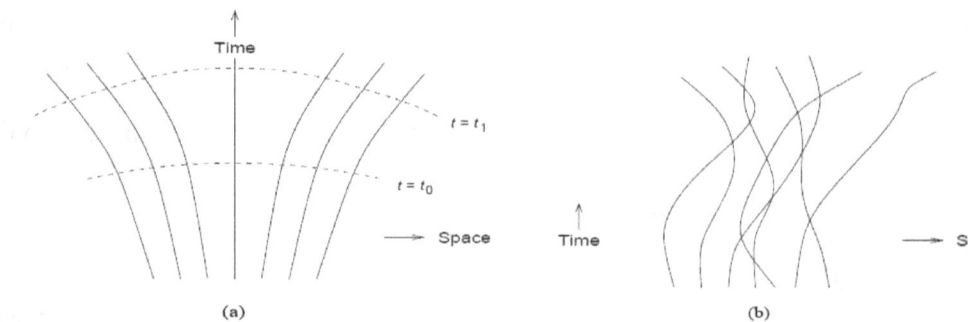

Figure 1. (a) Systematic large-scale motions of clusters; (b) arbitrary motions of small-scale objects (from Narlikar 2010, p. 230).

A note about the very important notion of substratum, so defined by Harwit: "The substratum in any cosmic model is a matrix of geometrical points all of which move in the idealized way required by the model" (2006, p. 480). Usually, it is utilized as a kind of perfectly continuous kinematic background. Thus we can look at the substratum as a geometer looks at a coordinate system. Interpreted in such a way the substratum "is nothing but a reference frame in uniform expansion" (Wegener 2000, p. 12).[20] However, we should not forget the nature of its points: each one of them is an entity ideally "containing" all the matter present in a given region of the universe. Therefore, the substratum – at a "large-scale level of abstraction", as it were – can be considered as an entity *in its own right*, whose particles are its real "atomic" (i.e., indivisible) material constituents.

2.2. WP's role at the foundations of standard cosmology

As I have already said WP is often completely neglected in most cosmology textbooks. Instead, it is the CP – stating that universe is spatially homogeneous and isotropic on large scales – that dominates the stage.

Standard cosmological models (the best description of the large-scale structure of our universe) are given by $< M . g_{ab} . T_{ab} >$, with M the spacetime manifold, g_{ab} the so-called *Friedmann-Lemaître-Robertson-Walker* (FLRW) *metric*, and T_{ab} the stress-energy tensor representing the

---

[20] Usually, this comoving frame is identified as the frame in which the cosmic microwave background radiation looks isotropic.

material contents (in the form of dust) of the universe. There are two ways whereby it is possible to derive the FLRW metric.[21]

In the first, and most used one, FLRW is derived just from CP. Spacetime is split up by imposing homogeneity and isotropy: isotropy guarantees that worldlines are orthogonal to each spatial hypersurface (Misner, Thorne and Wheeler 1973, p. 714), and cosmic time is a corollary of homogeneity (Rindler 2006, p. 359). WP is not explicitly mentioned.

In the second way – adopted, for example, by Bondi (1960), Narlikar (2010), Pauri (1995), Raychaudhuri (1979) – WP assumes a remarkably more elevated status: it is introduced first, it allows the definition of cosmic time and then the spacetime foliation. CP simply intervenes successively, imposing homogeneity and isotropy on spatial hypersurfaces.

Now, the really important point is that WP is *actually* always tacitly assumed in the first way as well. This fact is clearly shown by Rugh and Zinkernagel (2011).[22] The result is, in fact, that WP is mathematically disguised in the notion of isotropy.[23] So, one way or another, WP is *necessary* for a physically well-defined notion of cosmic time.[24] This significantly means that: "WP […] is a precondition for the CP; the former can be satisfied without the latter being satisfied but not vice versa" (2011, p. 417).

But another important difference between these two approaches to standard cosmological models can be highlighted by introducing the following two axiomatizations.

2.3. Deductive and constructive axiomatizations

A *deductive* axiomatic approach approximately "begins with a set of postulates concerning the existence of high level structures and/or principles and then proceeds by logical deduction to lower level phenomena which may be directly confronted by experiment" (Majer and Schmidt 1994, p. 17).

Many characteristics of this approach are shared by the so-called *top-down approach* to spacetime foundational questions, usually instantiated in

---

[21] See Mazzoni (1991).

[22] Also Pauri (1991, p. 334) hints at this thesis.

[23] Just to have an idea, see Wald's (1984, p. 92) technical formulation of CP, where the congruence of timelike curves is inserted into the definition of the isotropy.

[24] It is a necessary but *not* sufficient condition. One needs the further requirement that the congruence be orthogonal to the spatial hypersurfaces. However, this is questionable because some formulations of WP already include the orthogonality criterion (see Rugh and Zinkernagel 2011).

General Relativity with formulae like. "Spacetime *is* a 4-dimensional differentiable manifold... endowed with a semi-Riemannian metric...".[25] So higher level spatiotemporal structure is defined *from the outset* and is assumed as primitive and with a unifying explanatory role with respect to lower level structures (affine, projective, conformal) governing the physical behaviour of light and particles. In Coleman and Korté's (1994) opinion, modern spacetime realists have mainly taken such an approach.

The *constructive* (or *inductive*) axiomatic approach is the "reverse" of the deductive one: "The constructive axioms deal with directly observable phenomena at as low a level as possible. The aim is to formulate axioms which may be directly confronted by experiment, and then deduce from these low level axioms the existence of higher level structures" (Coleman and Korté 1994, p. 68). In particular, for spacetime theories: "The aim of a constructive axiomatic approach to a principle theory of spacetime structure is to exhibit the physical basis for the particular structural constraints which the principle theory postulates certain events must satisfy" (Coleman and Korté 2001, p. 257). Roughly, this view tries to clarify the physics behind the mathematics, namely, as regards spacetime, to trace its mathematical structures back to verifiable and measurable characteristics of few simple physical objects.

The problem of *deriving* the Riemannian spacetime of General Relativity by physically motivated axioms, rather than to *postulate* it at the outset, has been called by Castagnino (1971) the *inverse problem* of General Relativity. He proved that the *assumption* of a Riemannian spacetime geometry can be dispensed with: "If the spacetime paths of particles and light rays are experimentally known, [...] one can draw parallel lines, and construct an ideal geodesic clock that defines the metric over the whole manifold" (1971, p. 2203). In 1972, J. Ehlers, F. Pirani and A. Schild developed the most influential constructive approach to the general relativistic spacetime, called the *EPS approach*. Starting from an initial structureless set of point-events, and using only freely falling particles and light rays and a small set of experimentally verified constructive axioms, they were able to build up step by step all general relativistic spatiotemporal structures until reducing to the desired pseudo-Riemannian metric.

Now, I have introduced these views to stress that the CP-based approach can roughly be seen as an example of deductive methodology, and the WP-based approach as an example of the constructive one.[26] In the CP-based approach, in fact, homogeneity and isotropy select the FLRW metric from a general semi-Riemannian metric given from the outset. In the WP-

---

[25] As an example see Friedman (1983, p. 32).
[26] See Macchia (2011a, chapt. 3).

based approach, instead, we have an "inverse movement" *à la* Castagnino. Pauri explicitly states it: "Matters are turned around with respect to the standard approach: a geodesic is a geodesic of some metric; here a particular geodesic structure is assumed in order to *construct* a metric having certain desired properties" (1991, p. 320).[27]

2.4. Constructive axiomatization and relationism

That the EPS procedure complies with a relationist view concerning spacetime[28] is evident if one considers that EPS ontology consists only of particles and light rays, whereas spacetime, obtained solely from their characteristics, has been ontologically demoted to a "by-product".

Ehlers expressly underlines this:

It has been shown that on the basis of simple facts the spacetime geometry of General Relativity can be constructed without resorting to concepts or theorems of theories which presuppose such a geometry [...] Only concepts by which relations between events, particles and light rays are describable have been introduced. This fully agrees with Leibniz's position of viewing space and time not as objects but rather as sets of spatial or temporal relations among things. (quoted in Jammer 1993, p. 229)

In the cosmological context, the result is *supposedly* the same: the pure adoption of WP, with its assigning an ontological primacy to fundamental particles, constrains to an important element of *relationality* in the conceptual foundations of FLRW spacetime.

This possibility had already been stressed by J.G. Whitrow: "The three-dimensional spatial cross-section is determined solely by the fundamental particles, i.e. it is a relational space and not an absolute space with an independent existence of its own" (1980, p. 292).[29] More recently also Pauri has reached a similar conclusion: "The universal 'substratum' is defined by

---

[27] I am not claiming that this inverse approach in the cosmological context should necessarily be that of the EPS but that they could be closely related. On the other hand, EPS and WP particles are really similar: massive and freely falling point particles, gravitational monopoles, forming a congruence, spherically symmetric and non-rotating, with a unique timelike path, representing idealized observers each one with a clock necessary to parametrize its worldline. Also the modality to define time (the necessity of having a zero of time, an initial unit of time for all particles worldlines) is the same.

[28] I will not enter in slippery subtleties, here not crucial, concerning the form(s) of relationism involved.

[29] For a deeper analysis of Whitrow's view see North (1965, p. 366) and Macchia (2011a, par. 6.1).

a specific structure of *virtual* (not in the quantum-mechanical sense!) trajectories of *fundamental* particles which *relationally constitute* spacetime" (1991, p. 319).[30]

2.5. A second conclusion

From the WP-based approach it results that the nature of spacetime is supervenient on the substratum. Each spacetime point is identified with a fundamental particle. Hence substratum can be thought of as a space-*constituting* rather than as a space-*filling* set of particles (as instead happens in the CP-based approach).

In short: spacetime is *relationally* constituted by fundamental particles whose trajectories do *not* intersect.

3 Some final reconciling reflections

In the coming two subsections I highlight two insidious problems that the notion of worldline intersection can conceal. In the third subsection I dwell upon the worldlines of fundamental particles which, being non-intersecting, obviously do not face such problems. In the fourth and fifth subsections I will try to reconcile the preceding discordant conclusions of sect. 1.5 and 2.5.

3.1. Worldlines intersection and genidentity

A clarification needs to be made on the notion of coincidence. As Reichenbach (1928, p. 124) stresses, a coincidence is a concurrence of events at the same place and at the same time. A sort of simultaneity at the same place, even if, strictly speaking, is not a simultaneity of time points but an *identity*. In a coincidence, in the strict sense, position and time are identical for both events. Practically speaking, instead, such an identity *never* occurs since the two events involved, occupying the same spatiotemporal location, should be counted as one, and this does not make sense (furthermore, we could no longer individuate them). Only in *approximate* coincidences (e.g. two colliding spheres, two intersecting light rays) can this identity be realized. Therefore, the notion of coincidence calls

---

[30] Also Rugh and Zinkernagel (2009), through a detailed analysis of WP, obtain relational conclusions but confined only to cosmic time.

into question the notion of identity. The material counterpart of the logical notion of identity is what Reichenbach (and, before him, Kurt Lewin in 1922) calls *genidentity*, which is the "*physical identity* of a thing" (Reichenbach 1956, p. 38). So one can say that a physical thing is given by a series of events genidentical to each other. Different events are *states* of the same thing: for instance, its atoms of yesterday and its atoms of today are genidentical. Thus the points of a continuous time-like worldline are referred to as states of the *same* thing.

Now, if we look at the PCA paying attention to genidentity what results, given two intersecting worldlines of two different massive particles, is the unpleasant situation in which one has the "*same* point-event representing *two different identities*, i.e., even a logical contradiction!" (Pauri 2008, p. 176; italics are original). In such a case, as Pauri himself remarks, this fact contrasts with the PCA. His conclusions are really significant: "i) We cannot restrict the ontological bearing of the physical description of the world to its *theoretical structure*: we must take into account the *coordinative definitions* that link the latter to the *practice* of the experimentalist; ii) [...] the *effectiveness of the mathematical representation* of the *reduced ontology*[31] does not license anybody to operate a *reconstruction of the world* in terms of such ontology without being subject to philosophical *aporetic* consequences" (*ibid.*, p. 177; italics original).

Roughly speaking, Pauri is saying that these "theoretical" ontologies deduced from the physical-mathematical structures of our theories should be made "more concrete" by epistemic notions. Besides, he invites us to be cautious in our running about within these ontologies insofar as they always leave behind some "pieces of reality", or, even worse, they produce, when matched with the real world, aporetic results.

3.2. Worldlines intersection and ontology

Reichenbach (1928, p. 287) considers objective coincidences, on which all spatiotemporal order is based, as "physical events like any others", and whose "occurrence can be confirmed only within the context of theoretical investigation". "Since all happenings have until now been reducible to objective coincidences, we must consider it the most general empirical fact that the physical world is a system of coincidences". But, "what kind of physical occurrences are coincidences [...] is not uniquely determined by

---

[31] By "reduced ontology" Pauri means the *ontology of a theory*, for instance, of "material points" (Euler), or of "classical fields", of "elementary particles", "relativistic quantum fields", and so on (Pauri *ibid.*, p. 161).

203

empirical evidence", depending on "the totality of our theoretical knowledge". He gives (pp. 270-1) an illustration of this sort of *empirical underdetermination of coincidences* by taking up a continuous field like the electromagnetic one carried by *indistinguishable* particles like photons (see fig. 2). He says that both vertical lines and dotted slanted ones may be considered as the worldlines of the individual "field particles". So particle A_1 is genidentical with A_2, A_3, etc, *as well as* with B_2, C_3, D_4, etc. So, what counts as different points along a single field particle trajectory as opposed to neighboring points on different field particle trajectories? In a few words, the time-like worldlines of a continuous material field are not to be necessarily considered as striated in a definite direction, but there is a certain amount of arbitrariness in this choice.[32]

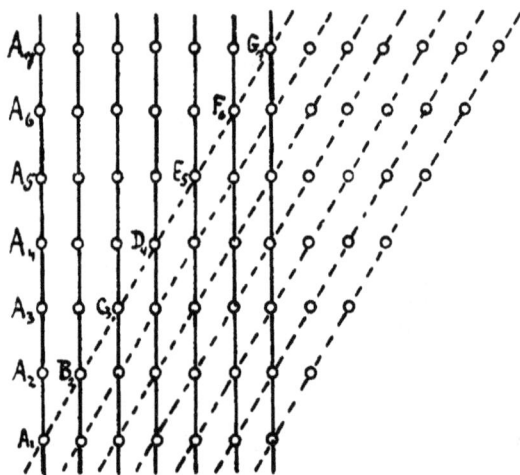

Figure 2. Arbitrariness of the striation of
worldlines in a continuous field (from
Reichenbach 1928, p. 271).

Thus, this empirical underdetermination of point coincidences induces an arbitrariness in the choice of the "grains" constituting the event ontology.

In the case of other material fields – Reichenbach continues – this problem does not arise. For instance, in the case of fields corresponding to "atomic matter" worldline bundles cannot be considered as arbitrary because there actually exists what Reichenbach calls a "natural striation" –

---

[32] Note that such a choice is arbitrary only within the time-like cones because the concept of genidentity cannot be satisfied by space-like worldlines.

"presumably because the costitutive particles are not indistinguishable from one another", Howard (1999, p. 491) comments. For "atomic matter", therefore, no arbitrariness exists in the choice of the "grains" of the manifold, and no problems about the preservation of genidentity now well defined.

This construction involving identical particles is attributed, by Reichenbach, to a paper by Einstein of 1920: *Äther und Relativitäts-Theorie*. Such an attribution – or rather: some consequences of Reichenbach's analysis ascribed by him to Einstein – is criticized by Howard (1999, p. 491). Without entering into the details, according to Howard the central point of Einstein's paper was arguing that the classical conception of the electromagnetic ether could be maintained (in relativity theory) only if one does not ascribe to it a definite state of motion, i.e. if one does not postulate the existence of a privileged frame of reference (the ether frame). So, this electromagnetic field, or other extended physical object fields "may not be thought of as consisting of particles that allow themselves to be tracked individually through time" (Einstein 1920, p. 10). In doing so, indeed, one would ascribe to such fields a definite state of motion, namely that of particle-like carriers of the field. The only acceptable way to avoid such an implicit attribution of a preferred state of motion to the electromagnetic field is a particle interpretation of this field in which the field particles are identical and indistinguishable. Therefore, this indistinguishability turns out to be necessary insofar as it implies an underdetermination in the ascription of worldlines to the particles (i.e., in the way particles worldlines lie within spacetime). And this means that no *unique* privileged state of motion can be tacitly associated with the field carried by these particles. It is such an underdetermination in the worldlines that induces, in turn, the underdetermination of their coincidences and then in the point coincidences (seen above) constituting the ontology.

Now, all this reasoning is not gratuitous. The empirical underdetermination of point coincidences "is not merely a trifling special case of a larger genus of empirical underdetermination. Instead, it is a fact about space-time event ontologies of fundamental significance for understanding the kinds of invariant structure that can live in a space-time" (Howard 1999, p. 493). In particular, here, it helps to pave the way for a better understanding of the peculiarities intrinsic to the WP-based ontological approach.

3.3. Cosmic substratum: the problems solver

Cosmological substratum is a sort of *quid* able to deftly avoid the preceding Reichenbachian "pitfalls". Indeed, substratum is considered as a *continuous* background whose fundamental particles, even if *conceptually* identical, are not indistinguishable from one another (each one represents a different region of the universe, each one has its own coordinate value). It is a sort of "atomic matter" field for which the concept of genidentity is well defined: substratum is an "aggregate" of genidentical worldlines insofar as each fundamental particle's worldline is a sequence of events characterized by a definite and unique identity. Such an identity is attached to the coordinatization, in the sense that each worldline is used to propagate the coordinate assigned to its particle from the arbitrary grid of space coordinates laid out on a spacelike hypersurface (see fig. 3). So, as it were, the coordinates are "carried", throughout the spacetime, by the worldlines. From this point of view it is natural, and really crucial!, that worldlines do *not* cross each other.[33] This is not a mere detail, but is the *physical basis* of WP itself.

Not by chance Narlikar (2002, p. 108) strongly underlines its necessity:

> It is worth emphasizing the importance of the non-intersecting nature of worldlines. If two galaxy worldlines did intersect, our coordinate system above would break down, for we would then have two different values of x^μ specifying the same point in spacetime (the point of intersection).[34]

It would be obviously absurd, therefore, for the same spacetime point-event to be described by two different and incompatible coordinates (namely by two different identifications).

In other words, at least in the FLRW models, there is the Reichenbachian "natural striation": the congruence of worldlines representing the average motion of matter is unique, that is, there is a preferred reference frame.[35] So, no point coincidence, no empirical

---

[33] On the other hand, to say that a family of worldlines forms a congruence means precisely that there is no crossing.

[34] x^μ (with $\mu = 1, 2, 3$) are the three space-like coordinates. So a typical worldline is $x^\mu = const$.

[35] In de Sitter's models, however, there is no unique choice of congruence (and this can have consequences on the deduced spacetime ontology, see Macchia 2011b). Note that the existence of a preferred reference frame does not obviously mean a unique coordinate representation: in fact, there are many different coordinate representations for the FLRW models (see Krasinski 1997, who outline at least five of these).

underdetermination, no arbitrariness in the "ontology grains", no problem with genidentity. Not bad!

But…

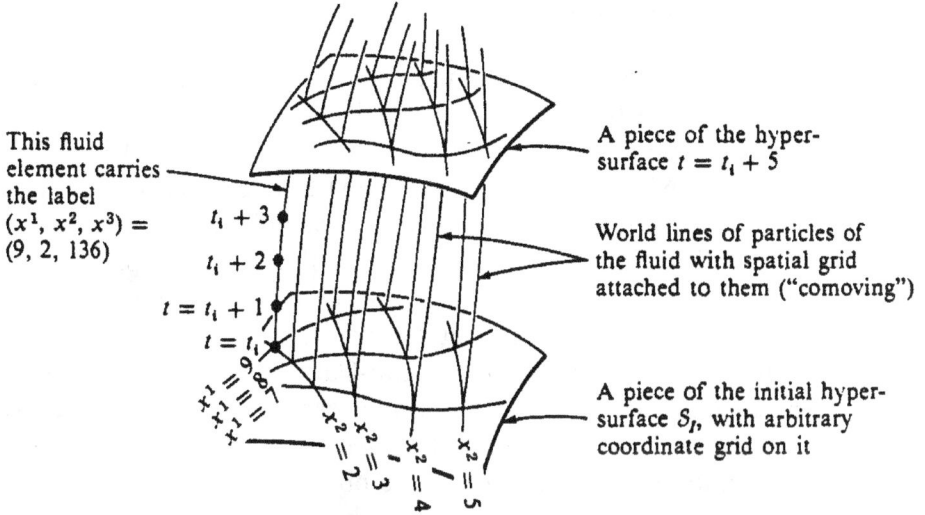

This fluid element carries the label $(x^1, x^2, x^3) = (9, 2, 136)$

A piece of the hyper-surface $t = t_i + 5$

World lines of particles of the fluid with spatial grid attached to them ("comoving")

A piece of the initial hyper-surface S_I, with arbitrary coordinate grid on it

Figure 3. Comoving synchronous coordinate system for the universe (from Misner, Thorne and Wheeler 1973, p. 716).

Let us come back to the PCA's dictates on the physical constitution of spacetime points.

Firstly, one can note that this non-intersecting character of fundamental particles' geodesics can be obviously read only ontologically, that is, it does not make any sense to think of fundamental particles as pointer-coincidences's (manqué) realizers, these particles being ideal point-particles, thus unobservable by definition.

Furthermore, it is evident how this necessary absence of any worldlines crossing jars with what has been stated by Einstein. Indeed, "the statement that they *do not* intersect" (Einstein, letter to Ehrenfest) is surely an invariant fact, therefore – according to the criterion of reality deriving from the PCA – strictly real. Consequently, from this perspective, no spacetime point would ever be able to "come into existence"!

To summarize: on the one hand, one obtains a spacetime *relationally* constituted by absolutely non-intersecting particles worldlines, on the other, a spacetime *relationally* constituted as well, but whose points acquire their only physically-objective contents *precisely* thanks to the worldlines' intersections. In other words, the same Einsteinian rejection in attaching a physical reality to a coordinatization, in the foundations of cosmology becomes a vital necessity.

3.4. No mystery, please…

Things, however, are not so mysterious. And the reasons should not be too difficult to infer from what I have said so far about the adoption of WP.

This principle embodies, in a sense, the reverse of that big conceptual passage from a Newtonian physical context to a general relativistic one. In this way WP, in its domain of validity, "blows up" the revolutionary general relativistic edification, PCA included. As Pauri (1991, p. 319) stresses: "Actually, the most important of the conceptual revolutions brought about by General Relativity is the dissolution of the universal chrono-geometrical 'substratum' provided by the Euclidean-Galilean framework". This means that in General Relativity "spacetime can no longer be, a-priori, a differentiable and metric manifold of *free-mobility* in which a universal notion of crono-geometrical *congruence* objectively resides" (*ibid.*). In other words, general relativistic spacetime, not having a pre-assigned geometry, is not characterized by a *group of movements* like, for instance, the group of rototranslations for the Euclidean 3-space, or the Poincaré group for the Minkowski spacetime.

The adoption of WP, instead, enables the reconstitution of that universal chrono-geometrical substratum. As in the Newtonian world, one obtains a universal and intrinsic notion of congruence, spatial and temporal. Actually, the notion of cosmic substratum – as we have seen in section 2.1 – allows us to establish a *quasi*-Newtonian time which is spatially and temporally global. "Quasi" because cosmic time is obviously not strictly Newtonian: its nature is not properly intrinsic and absolute, but contingent, insofar as it is determined by the properties of symmetries of the cosmic matter distribution; furthermore, cosmic time does not hold for every observer, in the sense that it is strictly measured only by the inertial fundamental observers belonging to the substratum.

Thus, in this "retrocession" to a quasi-Newtonian world, coordinates become again *physically meaningful parameters*. This fact is in line with what happens in other physical contexts. In general, indeed, physics itself requires us to adopt preferred coordinate systems. This is clearly stated by Ellis and Matravers (1995) who underline that, even if mathematical approaches to General Relativity insist that all coordinate systems are equal, what actually happens is that physicists, and in particular astrophysicists, often use preferred coordinate systems. The reason is that "some significant physical issues can *only* be sensibly tackled using (explicitly or implicitly) particular, well-adapted coordinate systems" (p. 778) (for instance, adapted to the symmetries of the system under study). This is not merely for calculational conveniences, but for understanding physical problems as well and defining the quantities of physical interest that enables us to understand such problems.

As regards astrophysics and cosmology, they explicitly say:

> There is a preferred rest frame and time coordinate in standard cosmology, and using any other coordinates simply obscures what is happening. The Cosmic Microwave Background Radiation determines the preferred rest frame (and associated time coordinate) to high accuracy. The dynamics of the standard model can be dealt with largely in a coordinate-free manner, but observational relations cannot. There are a small family of preferred spatial coordinates that focus either on spatial isotropy or homogeneity, and make observational analysis easy. The subject is completely opaque if other, ill-adapted coordinates are used. [...] In fact analyses of the growth of structure in the expanding universe, and the associated velocity flows, are mostly done in particular coordinates associated with taking a quasi-Newtonian approach to local astrophysics. (p. 781-782)

The only way to bridge the gap between "astrophysical practice" and "relativity ideology", as they call them, is a refinement of general covariance: "We do not abandon covariance of the theory, but move from general covariance (all coordinate systems are allowed, no matter how unsuitable) to restricted or physical covariance (the coordinates we use for physical applications exclude those that are wildly oscillating or are in other ways exceptionally badly adapted to the system at hand)" (p. 787).

What is involved in this situation is – they explain – "a type of symmetry breaking: the theory is covariant but the specific models we employ to understand various physical and astrophysical situations break that symmetry. The solutions of the equations that underlie the theory do not in general have all the symmetries of those equations. The family of physically useful solutions may have a smaller family of symmetries than the full set of solutions of the equations" (p. 785). Notwithstanding this, the results not based on a fully covariant approach *are* – they believe – still physically significant.[36]

---

[36] It should be noted, however, that Ellis and Matravers seem to be referring to a "passive version" of general covariance, i.e. invariance under general coordinate transformations, and not to an "active version", i.e. invariance under *active* diffeomorphisms (see footnote 6 in this paper). And physical content, typically, is ascribed to the active diffeomorphism invariance of General Relativity rather than to the passive one. Mathematically, these two types of transformations can be identified with one another. However, symmetry under the respective transformations can have different physical

On the other hand, coordinate systems and reference frames are not at all equivalent concepts, and general covariance states that all coordinate systems, *not* all reference frames, are equivalent. So that, in general, if a reference frame can be naturally associated with the actual movement of a system of bodies (as happens in FLRW models where the comoving frame is naturally associated to the divergent motions of clusters), the ability to perform a change of coordinates does not necessarily imply that such an association is still possible under the new coordinatization.

That being said, the situation instantiated by WP and PC is, in a certain sense, even more particular: they allow us to retrocede not only into a sort of pre-general relativistic ambit, but actually into a situation completely removed from the general relativistic context. Indeed, as Rindler (2006, p. 368) points out, FLRW metric applies "to *all* locally isotropic cosmological models, quite independently of General Relativity". That is, this metric can be found "without *any* of the assumptions of General Relativity" (*ibid.*): the simple and powerful assumptions of homogeneity and isotropy make it possible to deduce *kinematically* such a metric, without involving the dynamical approach of Einstein field equations (no energy-momentum tensor is needed).[37]

3.5. A last *tentative* conclusion

According to WP, all fundamental particles are freely falling. So, in this highly idealized picture one obtains a sort of global (all over the universe) cancellation – at very large scales – of the gravitational field and a consequent inertial frame as "large as the universe".[38] Such a global frame is composed of local inertial frames in which Special Relativity holds. In other words, cosmic substratum is locally compatible with a local flat Minkowskian spacetime. And in Minkowski spacetime – where there is no unique congruence of worldlines (i.e., no unique preferred frame) and no preferred cosmic time – there is no need to impose the non-crossing

---

significance. The problem is that there is no view, universally accepted, concerning what physical content is to be ascribed to the demand of general covariance.

[37] In a sense, such an independence may be seen as deriving from the fact that nothing in the Einstein field equations guarantees *a priori* that the circumstances described by WP and CP could actually verify in our universe.

[38] On why and how this is not in contrast with General Relativity see Bergia (1992, pp. 48-50).

criterion (as Rugh and Zinkernagel (2011, p. 418) remind us).[39] Therefore, it makes sense to claim a PCA-based ontology.

Now, the harmonic picture so far depicted breaks down as soon as we recall the gravitational effects due to the real masses associated with each fundamental particle. In these local *real* contexts the dictates of General Relativity necessarily emerge. That is to say, also from this most realistic viewpoint the possibility of a PCA-based ontology emerges once more.

In conclusion, it seems to me that the two relational ontologies analyzed in this paper are strictly dependent on the scales called into question by the underlying theoretical representations, which, in turn, depend on the way matter behaves in those domains. At very large scales, the regular expansion of the universe allows an idealization in which spacetime ontology depends on the fundamental particles; at small scales, the "chaotic" behavior of matter implies a spacetime ontology conforming to the PCA requirements.

By this I certainly do not pretend to obtain a "reconstruction of the world in terms of such ontolog[ies]" – to use Pauri's expression quoted at the end of sect. 3.1 – but, at least, this perspective seems to compose a unitary picture in which these two intrinsically contrasting relational ontologies are rendered compatible (though not universally valid), namely, not "subject to philosophical aporetic consequences" (*ibid.*).

References

Bergia, S. (1991): "Il Principio di Weyl, passo essenziale verso la precisazione della nozione di modello di universo". *Giornale di Astronomia* 1/2, pp. 48-53.

Bergia, S. (1992): "La nozione di tempo in relatività generale e in cosmologia", in F. Pollini and G. Tarozzi (eds.), *I concetti della fisica*. Modena, Mucchi, pp. 13-51.

Bergia, S. (1995): "Formulari, interpretazioni, ontologie: il caso delle teorie relativistiche", in G. Giuliani (ed.), *Ancora sul realismo*. Pavia, La Goliardica Pavese, pp. 47-68.

---

[39] On the other hand, in Minkowski spacetime there is no need for coordinates to be set up by the worldlines of the material constituents. But it is possible to do so, and in such a case the non-crossing criterion for the worldlines is once again necessary (see Peebles 1993, p. 249-51).

Bergia, S.; Mazzoni, L. (1999): "Genesis and Evolution of Weyl's Reflections on de Sitter's Universe", in H. Goenner et al. (eds.), *The Expanding Worlds of General Relativity*. Boston, Birkhäuser, pp. 325-342.

Bondi, H. (1960): *Cosmology* (2nd ed.). Cambridge, Cambridge University Press.

Castagnino, M.A. (1971): "The Riemannian Structure of Space-Time as a Consequence of a Measurement Method". *Journal of Mathematical Physics* 12, pp. 2203-2211.

Coleman, R.A.; Korté, H. (1994): "Constructive Realism", see Majer and Schmidt, pp. 67-81.

Coleman, R.A.; Korté, H. (2001): "Hermann Weyl: Mathematician, Physicist, Philosopher", in E. Scholz (ed.), *Hermann Weyl's* Raum-Zeit-Materie *and a General Introduction to His Scientific Work*. Basel, Birkhäuser Verlag, pp. 159-386.

Earman, J. (1989): *World Enough and Space-Time: Absolute Versus Relational Theories of Space and Time*. Cambridge, MIT Press.

Earman, J. (1995): *Bangs, Crunches, Whimpers, and Shrieks: Singularities and Acausalities in Relativistic Spacetimes*. New York, Oxford University Press.

Earman, J.; Norton, J. (1987), "What Price Space-Time Substantivalism? The hole story". *The British Journal for the Philosophy of Science* 38, pp. 515-525.

Ehlers, J. (1990): "Discussion", in B. Bertotti et al. (eds.), *Modern Cosmology in Retrospect*. Cambridge & New York, Cambridge University Press, pp. 29-30.

Ehlers, J. (2009): "Editorial Note to: H. Weyl, On the General Relativity Theory". *General Relativity and Gravitation* 41, pp. 1655-1660.

Ehlers, J.; Pirani, F.A.E.; Schild, A. (1972): "The Geometry of Free Fall and Light Propagation", in L. O' Raifeartaigh (ed.), *General Relativity – Papers in Honour of J.L. Synge*. Oxford, Clarendon Press, pp. 63-84.

Einstein, A. (1916) : "Die Grundlage der allgemeinen Relativitätstheorie". *Annalen der Physik* 49, pp. 769-822. Quotations are from the English translation: "The Foundation of the General Theory of Relativity", in H.A. Lorentz et al. (eds.), *The Principle of Relativity: A Collection of Original Memoirs on the Special and General Theory of Relativity*. London, Methuen, 1923; reprint: New York, Dover, 1952, pp. 109-164.

Einstein, A. (1920): *Äther und Relativitäts-Theorie. Rede gehalten am 5. Mai 1920 an der Reichs-Universität zu Leiden*. Berlin, Julius Springer.

Ellis, G.F.R.; Matravers, D. (1995): "General Covariance in General Relativity". *General Relativity and Gravitation* 27(7), pp. 777-788.

Ellis, G.F.R.; Williams, R. (2000): *Flat and Curved Space-Times*. Oxford, Oxford University Press.

Friedman, M. (1983): *Foundations of Space-Time Theories*. Princeton, Princeton University Press.

Goenner, H. (2001): "Weyl's Contributions to Cosmology", in E. Scholz (ed.), *Hermann Weyl's* Raum-Zeit-Materie *and a General Introduction to His Scientific Work*. Basel, Birkhäuser Verlag, pp. 105-137.

Harwit, M. (2006): *Astrophysical Concepts*. New York, Springer.

Howard, D. (1999): "Point Coincidences and Pointer Coincidences: Einstein on the Invariant Content of Space-Time Theories", in H. Goenner, J. Renn, J. Ritter, T. Sauer (eds.), *The Expanding Worlds of General Relativity* (Einstein Studies, vol. 7). Boston, Birkhäuser, pp. 463-500.

Jammer, M. (1993): *Concepts of Space. The History of Theories of Space in Physics*. New York, Dover Publications.

Kerszberg, P. (1986): "Le Principe de Weyl et l'invention d'une cosmologie non-statique". *Archive for History of Exact Sciences* 35, pp. 1-89.

Kerszberg, P. (1989): *The Invented Universe. The Einstein-de Sitter Controversy (1916-17) and the Rise of Relativistic Cosmology*. Oxford, Oxford University Press.

Lusanna, L.; Pauri, M. (2003): "General Covariance and the Objectivity of Space-Time Point-Events: The Physical Role of Gravitational and Gauge Degrees of Freedom in General Relativity". arXiv:gr-qc/0301040v1.

Macchia, G. (2011a): *Fondamenti della cosmologia e ontologia dello spaziotempo*. Urbino, PhD Thesis in Humanistic Sciences, University of Urbino.

Macchia, G. (2011b): "On the Relational Constitution of Cosmic Spacetime", in *Logic and Philosophy of Science, Proceedings of SILFS 2010 – International Conference of the Italian Society for Logic and Philosophy of Sciences*, December 15-17, 2010 Bergamo. Forthcoming.

Majer, U.; Schmidt, H.-J. (1994): *Semantical Aspects of Spacetime Theories*. Heidelberg, Spektrum Akademischer Verlag.

Mazzoni, L. (1991): *Dal postulato di Weyl al principio cosmologico*. Bologna, Dissertation in Physics, University of Bologna.

Misner, C.; Thorne, K.; Wheeler, J. (1973): *Gravitation*. San Francisco, Freeman.

Narlikar, J. (2002): *An Introduction to Cosmology*. Cambridge, Cambridge University Press.

Narlikar, J. (2010): *An Introduction to Relativity*. Cambridge, Cambridge University Press.

North, J.D. (1965): *The Measure of the Universe*. Oxford, Clarendon Press.

Norton, J.D. (1984): "How Einstein Found his Field Equations, 1912-1915". *Historical Studies in the Physical Sciences* 14, pp. 253-316. Reprinted in D. Howard e J. Stachel (eds.), *Einstein Studies*. Vol. 1. *Einstein and the History of General Relativity*. Basel, Birkhäuser, pp. 101-159.

Norton, J.D. (1987): "Einstein, the Hole Argument and the Reality of Space", in J. Forge (ed.), *Measurement, Realism and Objectivity*. Dordrecht, Reidel, pp. 153-188.

Norton, J.D. (1989): "Coordinates and Sovariance: Einstein's View of Space-Time and the Modern View". *Foundations of Physics* 19 (10), pp. 1215-1263.

Norton, J.D. (1993): "General Covariance and the Foundations of General Relativity: Eight Decades of Dispute". *Reports on Progress in Physics* 56(7), pp. 791-858.

Pauri, M. (1991): "The Universe as a Scientific Object", in E. Agazzi e A. Cordero (eds.), *Philosophy and the Origin and Evolution of the Universe*. Dordrecht, Kluwer A. P., pp. 291-339.

Pauri, M. (1995): "Spazio e tempo", in *Dizionario delle Scienze Fisiche*, Vol. V. Roma, Istituto della Enciclopedia Italiana (Treccani), pp. 433-464.

Pauri, M. (1996): "Oggettività e realtà", in F. Minazzi (ed.), *L'oggettività della conoscenza scientifica*. Milano, Franco Angeli, pp. 79-112.

Raychaudhuri, A.K. (1979): *Theoretical Cosmology*. Oxford, Clarendon Press.

Reichenbach, H. (1928): *Philosophie der Raum-Zeit-Lehre*. Berlin, Julius Springer. Page Numbers and Quotations from the English Translation: *The Philosophy of Space & Time*. New York, Dover, 1957.

Reichenbach, H. (1956): *The Direction of Time*, Berkeley. University of California Press. Page numbers and quotations from the 1971 edition.

Rindler, W. (2006): *Relativity. Special, General, and Cosmological*. Oxford, Oxford University Press.

Rovelli, C. (2004): *Quantum Gravity*. Cambridge, Cambridge University Press.

Rugh, S.; Zinkernagel, H. (2009): "On the Physical Basis of Cosmic Time". *Studies in History and Philosophy of Modern Physics* 40, pp. 1-19.

Rugh, S.; Zinkernagel, H. (2011): "Weyl's Principle, Cosmic Time and Quantum Fundamentalism", in D. Dieks et al. (eds.), *Explanation, Prediction, and Confirmation*. Dordrecht, Springer, pp. 411-424.

Stachel, J. (1989): "Einstein's Search for General Covariance, 1912-1915", in D. Howard and J. Stachel (eds.), *Einstein Studies*, Vol. 1. *Einstein and the History of General Relativity*. Basel, Birkhäuser, pp. 63-100. Based on a paper read at the Ninth International Conference on General Relativity and Gravitation, Jena, 1980.

Stachel, J. (1993): "The Meaning of General Covariance", in J. Earman, A. Janis, G. Massey (eds.), *Philosophical Problems of the Internal and External Worlds*. Pittsburgh, University of Pittsburgh Press, pp. 129-160.

Torretti, R. (1983): *Relativity and Geometry*. Oxford, Pergamon Press.

Wald, R. (1984): *General Relativity*. Chicago, The University of Chicago Press.

Wegener, M.T. (2000): "Ideas of Cosmology. A Philosopher's Synthesis", new version of a paper printed in Duffy, Wegener (eds.), *Recent Advances in Relativity Theory*, vol. 1, Hadronic Press, http://www.m-t-w.dk/2.%20COSMOLOGY/Ideas%20of%20Cosmology, %20A%20Philosopher.s%20Synthesis.pdf.

Weyl, H. (1922): *Space-Time-Matter*. New York, Dover Publications.

Weyl, H. (1930): "Redshift and Relativistic Cosmology". *Philosophical Magazine* 9, pp. 936-943.

Whitrow, G.J. (1980): *The Natural Philosophy of Time*. Oxford, Clarendon Press.

Dynamical Systems and the Direction of Time

Claudio Mazzola
University of Cagliari
mazzola.c@gmail.com

1 Introduction

Within the domain of philosophy of physics, the problem of the direction of time is commonly understood as the difficulty of accounting for the apparent directionality of macroscopic phenomena – such as those of thermodynamics or radiation – on the sole basis of time-reversible or time-symmetric equations – such as Newton's laws of motion or Maxwell's laws of electromagnetism.

Currently most accepted solutions to this problem (at least, outside quantum physics) typically rest on either (i) a de facto homogeneity in the initial distribution of matter in (our region of) the universe (e.g. Price, 1996) or (ii) on the high computational complexity of macroscopic systems (e.g. Hoover, 1999). Both solutions rest on a relational conception of time, according to which time is deprived of any intrinsic dynamical or directional property: in both cases, the question whether or not time is anisotropic – whether or not time is an arrow pointing toward a definite unique direction – is reduced to the question whether or not the dynamics of physical processes is forced to evolve in a unique temporal orientation.

This paper is dedicated to outline an alternative approach to the problem, centered on the possibility of providing time with intrinsic dynamical properties which depend on its own algebraic structure, therefore distinguishing between the directional properties of time from those of the physical processes taking place inside it. In particular, time shall be endowed with the algebraic properties of a monoid (i.e. a non-empty set together with an associative binary relation and an identity element), monoids being the least algebraic structures one needs to provide time with in order to speak of the dynamics of deterministic systems.

2 Dynamical Systems on Monoids

Arnold (1973) modeled deterministic systems on n-dimensional differentiable state spaces and governed by ordinary differential equations, such as those of classical mechanics and classical electromagnetism, by means of phase flows or continuous dynamical system, i.e. one-parameter groups of transformations indexed by the set R of time intervals, satisfying an identity and a composition requirement. More generally, deterministic systems on arbitrary non-empty state spaces and with (non-negative) integer or (non-negative) real time sets can be modeled by one-parameter families of transformations, indexed by $Z+$, Z, R or $R+$ (Giunti, 1997). Giunti and Mazzola (2012) further generalized this notion, allowing the time models of dynamical systems to consist of arbitrary monoids:

Definition 1. Dynamical System on a Monoid
A dynamical system on a monoid $L = (T, +)$ with identity 0, denoted by DS_L, is an ordered pair $DS_L = (M, (g^t)_{t \in T})$ such that
    6 M is a non-empty set,
    7 $(g^t)_{t \in T}$ is a family of functions on M, indexed by T,
    8 for any $x \in M$ and any $t, v \in T$

$$g^0(x) = x, \tag{2.1}$$
$$g^{t+v}(x) = g^t(g^v(x)). \tag{2.2}$$

M represents the *state space* (or *phase space*) of the system, T its *time set* and L its *time model*. Finally, for any $t \in T$, the function g^t is called a *state transition* of duration t: intuitively speaking, its role is that of mapping the state $x \in M$ the system displays at an arbitrary time to the unique state $g^t(x) \in M$ the state is in after an interval of duration t.

Dynamical systems on monoids with different state spaces and different time models may nevertheless describe the same dynamics. In that case, we claim them to be *isomorphic*.

<u>Definition 2. Isomorphism between Dynamical Systems</u>

Let $DS_{L1} = (M_1, (g^{t1})_{t\in T1})$ be a dynamical system on a monoid $L_1 = (T_1, +)$ and let $DS_{L2} = (M_2, (g^{t2})_{t\in T2})$ be a dynamical system on a monoid $L_2 = (T_2, \oplus)$. A function f is a ρ-isomorphism of DS_{L2} in DS_{L1} if and only if

7. $\rho: T_2 \rightarrow T_1$ is a monoid isomorphism of L_2 in L_1, and
8. $f: M_2 \rightarrow M_1$ is a bijection and, for all $x_2 \in M_2$ and all $t_2 \in T_2$
$f(g^{t2}(x_2)) = g^{\rho(t2)}(f(x_2))$.   (2.3)

<u>Definition 3. Isomorphic Dynamical Systems on Monoids</u>

Let DS_{L1} be a dynamical system on L_1 and let DS_{L2} be a dynamical system on L_2. DS_{L2} is isomorphic to DS_{L1} if and only if there exist f and ρ such that f is a ρ-isomorphism of DS_{L2} in DS_{L1}.

Isomorphism is an equivalence relation on any given set of dynamical systems on monoids; for this reason, isomorphic dynamical systems may be understood as being dynamically identical (Giunti and Mazzola, 2012).

2.1. Reversible Dynamics

Requiring the time model of a dynamical system to satisfy the sole algebraic properties of a monoid allows us to reduce sensitively the mathematical structure needed for describing the evolution of a deterministic system. This way, we are placed in a position to distinguish among a cluster of otherwise tangled notions of reversible dynamics, which the classical debate on time's arrow does not take into account.

The weakest requirement one would reasonably be willing to hold so that a dynamical system can be called *reversible* in the proper sense is that the system is capable of recovering any of its states, no matter what state transition it has undergone. Such a requirement is formally encoded in the following definition:

<u>Definition 4. Reversible Dynamical System</u>

A dynamical system $DS_L = (M, (g^t)_{t\in T})$ on a monoid $L = (T, +)$ is reversible if and only if for any $x \in M$, for any $t \in T$ there exists $r \in T$ such that
$g^r(g^t(x)) = x.$ (2.4)

By hypothesis, reversible dynamical systems possess no primitive states, for all their states may be reached by means of at least one non-identical state transition of non-zero duration; for this reason, we say that they possess no Gardens of Eden. Moreover, reversible dynamical systems do not possess any fixed point unless that point is static, i.e. isolated from all others points in the state space.

A stronger notion is that of *strict reversibility*, according to which, after having undergone a state transition of duration t, all states of a dynamical system can be recovered by means of the *same* (not necessarily unique) backward state transition of duration r.

Definition 6. Strictly Reversible Dynamical System

A dynamical system $DS_L = (M, (g^t)_{t \in T})$ on a monoid $L = (T, +)$ is strictly reversible if and only if for any $t \in T$, there exists $r \in T$ such that for any $x \in M$

$$g^r(g^t(x)) = x. \tag{2.5}$$

By definition all strictly reversible dynamical systems are reversible, while the converse does not generally hold.

An even stronger notion of reversibility is that of *time invertibility*, according to which the time model L of a dynamical system must possess the algebraic properties of a group.

Definition 6. Time-Invertible Dynamical System

A dynamical system DS_L on a monoid L is time-invertible if and only if L is a group.

It would be easy to show that time invertibility implies strict reversibility (and hence reversibility), as well as all the following types of reversible behavior:

Definition 7. Logically Reversible Dynamical System

A dynamical system DS_L on a monoid L is logically reversible if and only if all its state transitions are injective.

Definition 8. Dynamical System with Complete Past

A dynamical system DS_L on a monoid L has complete past if and only if all its state transitions are surjective.

Definition 9. Completely Logically Reversible Dynamical System

A dynamical system DS_L on a monoid L is completely logically reversible if and only if all its state transitions are bijective.

Logical reversibility may be understood as an epistemic form of reversibility: for, though logically reversible dynamical systems may not be reversible, any of their states may be retrieved by means of the sole knowledge of one point along its trajectory. Strict reversibility is sufficient for a system to be logically reversible, for if a state transition of duration t mapped different states x and z into a unique image y, then it could not be the case that the same state transition of duration r could lead y back to both x and z. On the other hand, reversibility neither implies nor is implied by logical reversibility.

Complete past ensures that any state of a dynamical system may be reached through all of its state transitions, though it may be the case that no state transition is capable of leading one or more states back along their trajectories. Conversely, reversible and strictly reversible dynamical systems may not have complete past, for reversibility and strict reversibility only ensure that any state of a dynamical system is the image of at least some, but not necessarily all, of its state transitions. Complete logical reversibility is obviously equivalent to logical reversibility together with complete past.

Finally, dynamical systems on monoids may exhibit the property of *time-symmetry*.

<u>Definition 10. Time-Symmetric Dynamical System</u>
A dynamical system $DS_L = (M, (g^t)_{t \in T})$ on a monoid $L = (T, +)$ is time-symmetric if and only if it is completely logically reversible and there exists a function $\sim M: \to M$, called *dynamical inversion*, such that for any $x \in M$ and any $t \in T$

$$\sim(g^t(\sim(x))) = (g^t)^{-1}(x). \tag{2.6}$$

Time-symmetry models what in the literature is commonly referred to as 'reversibility', 'time-reversal invariance' or 'time-reversal symmetry' (Lamb and Roberts, 1998). Clearly, time-symmetry requires a dynamical system to be completely logically reversible since it demands all state transitions to possess an inverse function; however, it is logically independent on all other forms of reversibility. This may be attributed to the fact that, in addition on depending on the logical properties of the state transitions of a dynamical system, time-symmetry also depends on the very features of its state space M (Horwich, 1987; Albert, 2000).

3 From the Algebra of Time to a Dynamics of Time

Definition 1 demands that the family of state transitions of a dynamical system on a monoid behaves as a left monoid action on its state space. On the other hand, any monoid $L = (T, +)$ can operate as a left monoid action on itself, so that it can be dynamically interpreted as a family of state transitions on T. This way, any monoid can be endowed with a dynamical system on its own called its time system.

Definition 11. Time System of a Monoid
The time system of a monoid $L = (T, +)$, denoted by $TS(L)$, is the ordered pair $TS(L) = (I, (\iota^t)_{t \in T})$ such that
9   $I = T$,
10   for all $t \in T$ and all $i \in I$
$\iota^t(i) = t + i$.   (3.1)

On the other hand, any dynamical system may be associated with a monoid, consisting of its state transitions along with the operation of function composition, and whose identity element is the state transition corresponding to the identity element of its time model.

Definition 12. Transition Algebra of a Dynamical System
Let $DS_L = (M, (g^t)_{t \in T})$ be a dynamical system on a monoid $L = (T, +)$. The *transition algebra* of DS_L, denoted by $TA(DS_L)$, is the ordered pair $TA(DS_L) = (H, *)$, where
3. $H = \{h : h = g^t \text{ for some } t \in T\}$ and
4. $*$ is the standard operation of function composition.

It is easy to prove[1] that the transition algebra $TA(TS(L))$ of the time system of any monoid L is itself a monoid. Moreover, $TA(TS(L))$ is isomorphic to L, while the time system $TS(TA(TS(L)))$ of the transition algebra $TA(TS(L))$ of any time system $TS(L)$ is isomorphic to $TS(L)$. This guarantees that time itself may be endowed with an internal dynamics which solely depends on its algebraic properties, and that no such property is lost while moving from its algebraic to its dynamical representation.

Following the milestone contributions of Reichenbach (1956), Mehlberg (1961) and Grünbaum (1964), philosophical investigations on the direction of time have typically been formulated in topological terms, focusing on whether or not time is isotropic, that is, topologically identical in both directions. In the light of the above results we can recover and

---
[1]   See Proposition 1 and Proposition 2 in the Appendix.

express the original, dynamical meaning of the question "does time flow in a unique direction?", basing the answer to such a question on the algebraic properties one needs to provide time with in order to describe of physical phenomena[2].

It is a surprising result that in the case of time systems all kinds of reversibility, except for logical reversibility and those depending on it, collapse on each other. This result is a consequence of the fact that reversibility of a time system is logically equivalent to the fact that all the elements of its monoid possess a left inverse, while their possessing a right inverse is equivalent to the time system's having complete past (Mazzola and Giunti, 2012). Reversibility, strict reversibility and complete past of a time system accordingly coincide with its time-invertibility. Claiming that time cannot run backwards is therefore tantamount to claiming that time is not a group while, conversely, requiring time to be a group is equivalent to requiring its dynamics to be reversible.

In addition, the dynamical features of each monoid L are predominant over those of the processes which are modeled by dynamical systems having L as a time model. In fact, reversible time systems (or strictly reversible time systems, or time systems with complete past, or completely logically reversible ones) invariably demand that their time models display the algebraic properties of a group, and hence that all dynamical systems on that monoid be time invertible. On the contrary, the time models of reversible dynamical systems (or strictly reversible dynamical systems, or dynamical systems with complete past) do not generally satisfy such properties, for reversibility, strict reversibility and complete past do not logically entail time-invertibility. So, while the dynamical models of irreversible processes cannot be endowed with reversible time, it may well be the case that the dynamical models of reversible (or strictly reversible, or logically reversible, ...) processes are endowed with irreversible time.

This way, the crucial question becomes what algebraic properties should one provide time with, in order to be capable of modeling time-symmetric phenomena such as those of classical mechanics or electromagnetism.

---

[2] Contrary to the tradition established by Reichenbach, Mehlberg and Grünbaum, this approach evidently presupposes a "substantivalist" understanding of time, according to which time can be endowed with intrinsic and irreducible properties on its own (in our case, of an algebraic kind). In this sense, the approach here outlined is rather along the lines of the "heretic" view of time initiated by Earman (1974).

4 Symmetry

Ismael and van Fraassen (2003) emphasize the role of symmetry in revealing superfluous theoretical structure – that is, multiple and equivalent representations for the same phenomenon. In their terminology, any theory is basically composed of a theoretical ontology, out of which metaphysically possible worlds are constructed, and of a physically possible world, which is selected among the metaphysical possibilities by means of laws. Finally, empirical interpretation of a theory provides the latter with qualitative properties, which may be understood as the epistemic bridge connecting the theory and the phenomena it is supposed to describe. Symmetries which are evidence for superfluous theoretical structure are precisely those which (i) depend on the laws of a theory and (ii) preserve its qualitative features.

Under this light, mathematical dynamical systems may be understood as theories, whose ontology is determined by their very general definition and whose laws are given by the specific form of their state transitions. Since we are treating dynamical systems as purely abstract entities on which no interpretation has yet been laid down, we may dismiss (ii) and restrict our analysis to those symmetries which depend on state transitions.

Physical time is ordinarily represented by a one-dimensional differentiable manifold (or as one dimension in a four-dimensional manifold) which is diffeomorphic to the real line. As such, it is typically endowed with the algebraic structure of a group, together with commutativity and a linear order. It is easy to prove such a rich mathematical structure to possess an internal symmetry. In fact, the time system of any commutative group $L = (T, +)$ with identity 0 is naturally endowed with a dynamical inversion function , which coincides with the automorphism on T mapping any element of T to its algebraic inverse[3]. In general, automorphisms define the internal symmetries of the mathematical objects to which they apply (Weyl, 1952). The same is true, in the case of any arbitrary linearly ordered commutative group $L = (T, +, \leq)$, for the function $\sim$ on the state space of the corresponding time system. Let $L^+ = (T^+; +|_{T^+}, \leq|_{T^+})$ and $L^- = (T^-; +|_{T^-}, \leq|_{T^-})$ be two submonoids of L such that $T^+ = \{t \in T: 0 \leq t\}$ is the "positive" part of T according to the linear order $\leq$ and $T^- = \{t \in T: t \leq 0\}$ is the "negative" part according to $\leq$; then $\sim$ is a $\sim$ isomorphism between the time systems $TS(L^+)$ and $TS(L^-)$ of L^+ and L^- inverting their time order[4]. In the case under consideration, this means that transforms orbits of $TS(L)$ into dynamically indistinguishable orbits of

---

[3] See Proposition 3 in the Appendix.

[4] See Proposition 4 and Corollary 4.1 in the Appendix. Proposition 4 was suggested by Marco Giunti.

TS(L). Therefore, representing time as a linearly ordered commutative group is endowing the latter with redundant internal dynamics, for any displacement in time is at least represented twice: as a time-transition ι^t: $i \rightarrow \iota^t(i)$ and as a symmetrical time transition ι^{-t}: $\sim(i) \rightarrow \iota^{-t}(\sim(i))$.

This result can be further generalized to all time invertible dynamical systems on commutative linearly ordered groups. Let $L = (T, +)$ be a monoid possessing a nontrivial submonoid $L|_S = (S, +|_S)$, with $S \subset T$; then for any dynamical system $DS_L = (M, (g^t)_{t \in T})$ on M it is possible to define a dynamical system $DS_{L|S} = (M, (g^t|_S)_{t \in S})$ on $L|_S$, whose dynamics is a proper part of that of DS_L and which is called a *temporal section* of the latter. Linearly ordered groups naturally come provided with the two non-trivial submonoids consisting of their positive and negative parts; hence, for each time invertible dynamical system it is possible to define a pair of *chiral* temporal sections DS_{L+} and DS_{L-}, whose time models are the positive and the negative parts of the given one. Chiral temporal sections of time-symmetric dynamical systems on linearly ordered commutative monoids are always isomorphic[5]. Hence, for any transition $g^t: x \rightarrow g^t(x)$ taking place in the positive temporal section DS_{L+} of DS_L there exists a transition g^{-t}: $\sim(x) \rightarrow g^{-t}(\sim(x))$ being its exact negative duplicate in DS_{L-}, and vice-versa.

Following Ismael and van Frassen's suggestion, we may look at this dynamical redundancy as an evidence – though still not a proof – that standard modeling of time suffers of an excess in algebraic structure. If that was really the case, then it would be possible to discard part of that structure, reducing it to that of a monoid, and to describe time-symmetric deterministic processes with the aid of irreversible time models. Of course, one might continue in modeling time as a group for ease of calculation; however, the apparent reversibility of such a time model would be the mere product of a mathematical artifact without any physical meaning.

5 Conclusion

Focusing on the algebraic properties of time allowed us to provide formal systems useful to express its internal dynamics. In addition, we pointed out that the sole way for such dynamics to be reversible in a non purely logical sense is by shaping time as a group. Finally, we proved that representing time as a group, together with some minor constraints such as commutativity and linear ordering, is sufficient for doubling the trajectories of any time-symmetric or time-reversal invariant dynamical system. In the light of these results, we can draw two main conclusions: (a) that, contrary

---

[5] See Proposition 5 in the Appendix.

to the received view, the time-reversal invariance is neither necessary nor sufficient for making time reversible, for time symmetry is logically independent on time invertibility and (b) that adding time invertibility to time-symmetric dynamical systems may generate superfluous dynamics. Both conclusions call for a reexamination of the standard philosophical attitude concerning the problem of the direction of physical time, and for a deeper investigation on the least mathematical structure needed to model macroscopic physical phenomena[6].

Appendix: Propositions and Proofs

Proposition 1. The transition algebra TA(TS(L)) of the time system TS(L) of a monoid L is a monoid isomorphic to L.

Proof.
Let $TS(L) = (I, (\iota^t)_{t \in T})$ be the time system on a monoid $L = (T, +)$ with identity 0 and let $TA(TS(L)) = (H, *)$ be the transition algebra of $TS(L)$. Proof of Proposition 4 will proceed in two steps. First, we shall prove that $TA(TS(L))$ is a monoid with identity ι^0; second, we shall prove that such monoid is isomorphic to L.

To prove that $TA(TS(L))$ is a monoid with identity ι^0, it is sufficient to notice that

5. $TA(TS(L))$ is closed with respect to the composition rule $*$: for any $\iota^t, \iota^v \in H, \iota^t * {}^v$ $= \iota^{t+v} \in H$;

6. $*$ is associative: associativity is a general property of the operation of function composition;

7. $\iota^0 \in H$ and, for any $h \in H$ and $t \in T$ such that $h = \iota^t, \iota^0 * \iota^t = \iota^{0+t} = \iota^t = \iota^{t+0} = \iota^t * \iota^0$.

To prove that $TA(TS(L))$ is isomorphic to L, let $\rho: T \to H$ be the family $(\iota^t)_{t \in T}$. Then:

4. ρ maps identity element into identity element: $\rho(0) = \iota^0$;

5. ρ is structure-preserving: for any $t, v \in T: \rho(t + v) = \iota^{t+v} = \iota^t * \iota^v$;

6. ρ is bijective:

    a. ρ is injective: for any $t, v \in T$,

        $t \neq v$

        $t + 0 \neq v + 0$

        $\iota^t(0) \neq \iota^v(0)$

        $\iota^t \neq \iota^v$

        $\rho(t) \neq \rho(v)$;   (A1)

    b. ρ is surjective: by Definition 12, for any $h \in H, h = \iota^t = \rho(t)$ for some $t \in T$.

□

Lemma 1. Let L_1 be a monoid with time system $TS(L_1)$ and let L_2 be a monoid with time system $TS(L_2)$. Any monoid isomorphism ρ of L_2 in L_1 is a ρ-isomorphism of $TS(L_2)$ in $TS(L_1)$.

---

[6]   Grateful acknowledgments to Marco Giunti and to an anonymous referee for their fruitful comments.

Proof.
Let $L_1 = (T_1, +)$ be a monoid with time system $TS(L_1)$ and let $L_2 = (T_2, \oplus)$ be a monoid with time system $TS(L_2)$. Let $\rho: T_2 \to T_1$ be a monoid isomorphism of L_2 in L_1. Hence, for any $t_2 \in T_2$ and any $i_2 \in I_2 = T_2, \rho(\iota^{t_2}(i_2)) = \rho(t_2 \oplus i_2) = \rho(t_2) + \rho(i_2) = \iota^{\rho(t_2)}(\rho(i_2))$. (A2)

Hence, according to Definition 2, ρ is a ρ-isomorphism between time systems. □

Proposition 2. *For any time system $TS(L)$, $TS(TA(TS(L)))$ is isomorphic to $TS(L)$.*

Proof.
Let $TS(L)$ be the time system of a monoid L, let $TA(TS(L))$ be the transition algebra of $TS(L)$ and let $TS(TA(TS(L)))$ be the time system of $TA(TS(L))$. According to Proposition 1, $TA(TS(L))$ is isomorphic to L. Hence, by Lemma 1 and Definition 3, $TS(TA(TS(L)))$ $TS(L)$ are isomorphic. □

Lemma 2. *Let $TS(L) = (I, (\iota^t)_{t \in T})$ be the time system of a commutative group L. Then the function $\sim: T \to T$ mapping any $t \in T$ to its algebraic inverse is a dynamical inversion function on I.*

Proof.
Let $TS(L) = (I, (\iota^t)_{t \in T})$ be the time system of a commutative group L and let $\sim: T \to T$ be the function such that, for any $t \in T$, $\sim(t) = -t$. Then, for any $t \in T$ and any $i \in I = T$,

$(t-i) + (\sim(t-i)) = 0$
$-i + (\sim(t-i)) = -t$
$\sim(t-i) = -i + (-t);$ (A3)
hence, by commutativity:
$\sim(\iota^t(\sim(i))) = \sim(\iota^t(-i)) = \sim(t-i) = -i + (-t) = -t + (-i) = (\iota^t)^{-1}(i).$ (A4)
So, by Definition 10, $\sim$ is a dynamical inversion function on I. □

Proposition 3. *All time-invertible time systems on commutative monoids are time-symmetric.*

Proof.
Let $TS(L) = (I, (\iota^t)_{t \in T})$ be the time system of a commutative group L. Being time-invertible, $TS(L)$ is also completely logically reversible. Hence, given Lemma 2 and Definition 10, $TS(L)$ is time-symmetric. □

Lemma 3. *Let $L = (T, +, \leq)$ be a group with identity 0 and let L^+ and L^- be its positive and negative parts; then L^+ and L^- are linearly ordered submonoids of L.*

Proof.
Let $L = (T, +, \leq)$ be a group with identity 0 and let L^+ and L^- be its positive and negative parts; we shall prove that L^+ is a linearly ordered submonoid of L, the same proof applying *mutatis mutandis* for L^-:

6. $T^+ \subseteq T$: by hypothesis;

7. $0 \in T^+$ by hypothesis;
8. $+|_{T+}$: by inheritance from $+$;
9. T^+ is closed under $+|_{T+}$:
    a. if there existed $t, v \in T^+ - \{0\}$ such that $t + v \notin T^+ - \{0\}$, then $t+v \in T^- - \{0\}$ and therefore

$t + v \leq t$	and	$t + v \leq v,$
$-t + (t + v) + (-v) \leq -t + t + (-v)$	and	$-t + (t + v) + (-v) \leq -t + v + (-v),$
$0 \leq -v$	and	$0 \leq -t,$
$v \leq 0$	and	$t \leq 0,$ (A5)

    and hence $t, v \notin T^+$, against the hypothesis;
    β. if $t, v \in T^+$ and $t = 0$ or $v = 0$, then closure holds trivially.   ☐

Proposition 4. Let $L = (T, +, \leq)$ be a linearly ordered group with identity 0 and let L^+ and L^- be its positive and negative parts; then the function $\sim$: $T^+ \to T^-$ such that, for any $t \in T^+$, $\sim(t) = -t$, is

1. *an isomorphism from $(T^+, \leq|_{T+})$ to $(T^-, \leq|_{T-})$,*
2. *an isomorphism from $(T^+, +|_{T+})$ to $(T^-, +|_{T-})$ if and only if $(T, +)$ is commutative.*

Proof.
Let $L = (T, +, \leq)$ be a linearly ordered group with identity 0, let L^+ and L^- be its positive and negative parts and let $\sim$: $T^+ \to T^-$ such that, for any $t \in T^+$, $\sim(t) = -t$. By Lemma 3, we know that L^+ and L^- are both linearly ordered monoids.

To prove statement 1, it is then sufficient to notice that:

10. $\sim$ is bijective: by hypothesis, any $t \in T^+$ has a unique algebraic inverse $-t = \sim(t)$;
11. for any $t, v \in T^+$

$$t \leq v$$
$$\sim(t) + t + \sim(v) \leq \sim(t) + v + \sim(v)$$
$$\sim(v) \leq \sim(t). \tag{A6}$$

To prove statement 2, it is then sufficient to notice that:

12. $\sim$ is bijective: as before;
13. $\sim$ maps identity element into identity element: $\sim(0) = -0 = 0$;
14. $\sim$ is structure-preserving if and only if L is commutative:
    a. if L is commutative, then for any $t, v \in T^+$

$$\sim(t + v) = -(t + v) = -v + (-t) = -v + -(t) \tag{A7}$$

    b. if L is not commutative, then for some $t+v \in T^+$

$$t + v \neq v + t$$
$$-(t + v) + (t + v) + (-(v + t)) \neq -(t + v) + v + t (-(v + t))$$
$$-(v + t) \neq -(t + v)$$
$$\sim(v + t) \neq -(t + v) = -v + (-t) = \sim(v) + (\sim(t)). \tag{A8}$$

Corollary 4.1. Let $L = (T, +, \leq)$ be a linearly ordered commutative group with identity 0 and let L^+ and L^- be its positive and negative parts with time systems $TS(L^+)$ and $TS(L^-)$ respectively; then the function $\sim: T^+ \to T^-$ such that, for any $t \in T^+$, $\sim(t) = -t$, is a $\sim$-iso-morphism of $TS(L^+)$ in $TS(L^-)$.

Proof.
Let $L = (T, +, \leq)$ be a linearly ordered commutative group with identity 0, let L^+ and L^- be its positive and negative parts, let $TS(L^+)$ and $TS(L^-)$ be their time systems and let $\sim: T^+ \to T^-$ be the function such that, for any $t \in T^+$, $\sim(t) = -t$. By Proposition 4, $\sim$ is a monoid isomorphism of L^+ in L^-; hence, by Lemma 1, $\sim$ is a $\sim$-isomorphism of $TS(L^+)$ in $TS(L^-)$. ☐

Proposition 5. *Let $DS_L = (M, (g^t)_{t \in T})$ be a dynamical system on a commutative linearly ordered group $L = (T, +, \leq)$ and let DS_{L+} and DS_{L-} be temporal sections of DS_L on L^+ and L^- respectively. If DS_L is time symmetric, then the dynamical inversion function $\sim$ on M is a ρ-isomorphism of DS_{L+} in DS_{L-}.*

Proof.
Let $DS_L = (M, (g^t)_{t \in T})$ be a dynamical system on a commutative linearly ordered group $L = (T, +, \leq)$, let DS_{L+} and DS_{L-} be temporal sections of DS_L on L^+ and L^-, let $\rho: T^+ \to T^-$ be the function such that, for any $t \in T^+$, $\rho(t) = -t$ and let $\sim$ be the dynamical inversion function on M. Then, for any $x \in M$ and any $t \in T^+$:

$\sim(g^t(\sim(x)) = g^{-t}(x)$
$g^t(\sim(x)) = \sim(g^{-t}(x))$
$g^{\rho(-t)}(\sim(x)) = \sim(g^{-t}(x))$.

Accordingly, by Definition 2, $\sim$ is a ρ-isomorphism of DS_{L+} in DS_{L-}. ☐

References

Albert, D. (2000): *Time and Chance*. Cambridge and London, Harvard University Press.

Earman, J. (1974): An Attempt to Add a Little Direction to 'The Problem of the Direction of Time'. *Philosophy of Science*, 41(1): 15-47.

Giunti, M. (1997): *Computation, Dynamics and Cognition*. New York and Oxford, Oxford University Press.

Giunti, M.; Mazzola, C. (2012): "Dynamical Systems on Monoids: Toward a General Theory of Deterministic Systems and Motion", forthcoming in G. Minati, M. Abram and E. Pessa, *Methods, Models, Simulations and Approaches Towards a General Theory of Change. Proceedings of the Fourth National Conference of the Italian Systems Society*. World Scientific, Singapore.

Grünbaum, A. (1964): *Philosophical Problems of Space and Time*, Studies in the Philosophy of Science. Dordrecht, Reidel, second enlarged edition, 1973.

Hoover, W.G. (1999): *Time Reversibility, Computer Simulation and Chaos*. Singapore, World Scientific.

Horwich, P. (1987): *Asymmetries in Time. Problems in the Philosophy of Science*. Cambridge, MIT Press.

Ismael, J.; van Fraassen, B. (2003): "Symmetry as a Guide to Superfluous Theoretical Structure", in K. Brading and E. Castellani (eds.), *Symmetries in Physics. Philosophical Reflections*. Cambridge and New York, Cambridge University Press, pp. 371–392.

Lamb, J.S.; Roberts, J. A. (1998): "Time-Reversal Symmetry in Dynamical Systems: a Survey". *Physica D: Nonlinear Phenomena*, 112: Proceedings of the Workshop on Time-Reversal Symmetry in Dynamical Systems, pp. 1–39.

Mazzola, C.; Guintini, M. (2012): "Reversible Dynamics and the Directionality of Time", forthcoming in G. Minati, M. Abram and E. Pessa, *Methods, Models, Simulations and Approaches Towards a General Theory of Change. Proceedings of the Fourth National Conference of the Italian Systems Society*. World Scientific, Singapore.

Price, H. (1996): *Time's Arrow and Archimedes' Point. New Directions for the Physics of Time*. New York and Oxford, Oxford University Press.

Reichenbach, H. (1956): *The Direction of Time*. Berkeley and Los Angeles, The University of California Press.

Reichenbach, H. (1958): *The Philosophy of Space and Time*. Berkeley and Los Angeles, The University of California Press.

Weyl, H. (1952): *Symmetry*. Princeton, Princeton University Press.

Poincaré's physics
'Conventionalism' and the dispute regarding
the Relativity priority

Giulia Giannini
University of Bergamo
giulia.giannini@unibg.it

1 Introduction. The dispute regarding Relativity priority

In 1953, Edmund T. Whittaker (1873-1956) dedicated a chapter of the second volume of his *A History of the Theories of Aether and Electricity* to the relativity theory of Poincaré and Lorentz[1]. In its title Einstein's name was completely omitted and a minimal importance was given to his article *On the Electrodynamics of Moving Bodies*.

In his study 1921 *Relativitätstheorie*[2] Wolfgang Pauli (1900-1958) had already revalued the role of Poincaré in the history of special relativity but it was only with Whittacker's book that a controversial debate arose regarding priority that involved, among others, Max Born (1882-1970)[3], Gerald Holton[4], Marie-Antoniette Tonnelat[5], Arthur I. Miller[6], Abraham Pais[7]...

[1] Whittaker (1951-1953).
[2] Pauli, (1921).
[3] See: Einstein, Born, M., Born, H., (1969) and Born, M., (1964).
[4] Holton (1988).
[5] Tonnelat (1971).
[6] Miller (1973, 1981, 1996).

Even though critics agreed on the presence of all mathematical elements in Poincaré's works, they disagreed on the existence in them of a real "relativistic spirit". In particular, many critics judged Poincaré's philosophy, usually identified with the term 'conventionalism', as an 'epistemological obstacle' in the formulation of a theory. To them, Poincaré's epistemology implied a conservative attitude that did not allow the formulation of a real theory, whatever it was. According to this point of view, although Poincaré achieved to develop all mathematical aspects of special relativity, he was unable to recognize its epistemological consequences. In this sense, the dispute regarding relativity priority would loose its meaning and Einstein would be considered the only real discoverer of the theory.[7]

2 The aim: a revaluation of conventionalism

The first goal of this study is to show the necessity of a revaluation of Poincaré's epistemological perspective. An analysis of his physical work can throw new light on his epistemological attitude and, consequently, on his contribution to relativity theory. This paper will not deal with questions related to the formalism that are already widely discussed in literature that agree to find in Poincaré's works all the mathematical basis of the theory. Instead it will show insubstantiality of the bachelardian concept of 'epistemological obstacle' with regards to Poincaré's contribution in special relativity.

This aim will firstly be achieved by showing that if, on the one hand, the term 'conventionalism' can be appropriately used to designate Poincaré's approach to Geometry then on the other hand, his position about Physics is more complex. In particular this analysis will focus on his criticism of Mechanics, on the importance he placed on experimental confirmation and on the operational origin of physical concepts and theories.

3 The criticism of Mechanics

By 1880, Poincaré contributed to the criticism of Mechanism that characterized the origin of different world pictures in the second part of the 19th Century. Indeed, at this time it was possible to witness the clash of different theoretical positions concerning distinct world views. Mechanism had been the dominant paradigm over two centuries. Now, however, there

---

[7] Pais (1982).

were new attempts to formulate unified world pictures, based on rising physical disciplines, such as Thermodynamics (its first two principles date back to the 1860s) and Maxwell's electromagnetism (1873). The electromagnetic world picture, in opposition to the mechanistic one, tried to explain all the natural phenomena, not by reducing them to matter and motion, but rather through the laws of the electromagnetic field. The aim of such a view about Nature was to base all physics on Electromagnetism, which was conceived as the basic discipline to which all the others had to be reduced.

Poincaré's criticism of Mechanism, already present in the afterword of Leibniz's Monadologie (1880) edited by Emile Boutroux[8], appears as a constant element in his works. At the beginning, it was based on the impossibility of finding a unified mechanical explanation of phenomena and on the problems caused by the relationship between Mechanism and the new experimental evidences. Poincaré claimed that the existence of one mechanical model implied the existence of an infinite number of mechanical models.

Therefore, it was impossible to determine, among the infinite possible mechanical models, which would be most suitable for describing natural phenomena.

Neither the experience nor the convenience (used, for example, in geometry) could help in the choice among the different mechanical models: so, such a choice was founded on purely subjective and metaphysical considerations. Due to the impossibility of defining a single mechanical model, Poincaré argued for their insubstantiality. The infinity of such models was, in fact, the first step towards their loss of meaning. Moreover, Poincaré stated, as early as 1894, the uselessness of research into a mechanical model. For him, it was not necessary to find a mechanical explication, but rather to look for unity of nature, namely for the common features of all the theories.

Starting from his lessons in 1887-1888, such a unity appeared as the only aim of scientific research. In his paper *Les relations entre la physique expérimentale et la physique mathématique* (1900)[9] Poincaré declared that the attempt to find a unitary view of nature clashed with the difficulties linked to the mechanistic interpretation of electrical phenomena. Then, in 1893, he showed that the mechanical effort of giving a unitary explanation to all phenomena by means of mass and motion met with several obstacles. The physicists had difficulties reconciling mechanical description with experimental data. In particular, such an attempt proved to be incompatible

---

[8] Poincaré (1880).
[9] Poincaré (1900a).

with phenomenal irreversibility. The experience showed a number of irreversible events whereas mechanist hypothesis presupposed the reversibility of all phenomena. Thus, the aim of finding unity of nature, while essential, could not be pursued in a mechanist way.

4 The operational origin of physical concepts and theories

Poincaré's general criticism of Mechanism was followed, from at least 1895, by a deep criticism of some distinctive concepts of Mechanics. By an analysis of his principal notions, e.g. the idea of mass or absolute concepts of time and space, Poincaré attacks classical Mechanics at its foundations, showing its implicit assumptions and hidden contradictions.

By 1895, in *A propos de la théorie de Larmor*, Poincaré affirmed the impossibility of observing absolute motion[10]. Later, in *La mesure du temps* (1898), he showed the conventional character of measuring procedures of temporal intervals and, more generally, conventionality of time:

> Nous n'avons pas l'intuition directe de l'égalité de deux intervalles de temps. Les personnes qui croient posséder cette intuition sont dupes d'une illusion [...].La simultanéité de deux événements, ou l'ordre de leur succession, l'égalité de deux durées, doivent être définies de telle sorte que l'énoncé des lois naturelles soit aussi simple que possible. En d'autre termes, toutes ces règles, toutes ces définitions ne sont que le fruit d'un opportunisme inconscient.[11]

In another paper, published in honour of the jubilee of Lorentz's doctoral thesis[12], Poincaré introduced his method of clock synchronisation by light signals. Then, through a physical interpretation of Lorentz's local time, he reaffirmed the inexistence of Absolute Time. Starting from this text Poincaré also introduced the "principle of relative motion". Furthermore in a lecture on the principles of mechanics, again in 1900, he claimed:

> 1° Il n'y a pas d'espace absolu et nous ne concevons que des mouvements relatifs; cependant on énonce le plus souvent les faits mécaniques comme s'il y avait un espace absolu auquel on pourrait les rapporter;
> 2° Il n'y a pas de temps absolu; dire que deux durées sont égales, c'est une assertion qui n'a par elle-même aucun sens et qui n'en peut acquérir un que par convention;

---

[10] Poincaré (1895, 412).
[11] Poincaré (1898).
[12] Poincaré (1900b).

3° Non seulement nous n'avons pas l'intuition directe de l'égalité de deux durées, mais nous n'avons même pas celle de la simultanéité de deux événements qui se produisent sur des théâtres différents; [...]

4° Enfin notre géométrie euclidienne n'est elle-même qu'un sorte de convention de langage; nous pourrions énoncer les faits mécaniques en les rapportant à un espace non euclidien qui serait un repère moins commode, mais tout aussi légitime que notre espace ordinaire; l'énoncé deviendrait ainsi beaucoup plus compliqué; mais il resterait possible.

Ainsi l'espace absolu, le temps absolu, la géométrie même ne sont pas des conditions qui s'imposent à la mécanique; toutes ces choses ne préexistent pas plus à la mécanique que la langue française ne préexiste logiquement aux vérités que l'on exprime en français.[13]

For Poincaré the concepts of Absolute Space, Absolute Motion and Absolute Time were already meaningless within Classical Mechanics[14]. The impossibility of determining them in an experimental way showed, in Poincaré's opinion, that they were empty notions, alien to physical processes[15]. Poincaré's criticism also took into account the concept of Mass. Related to the electromagnetic field, Mass depended on direction and velocity of body motion and makes sense only for slower than light velocities. In several papers, notably the 1904 Saint-Louis lecture and *La fin de la matière* (published in 1906 and since 1907 included in *La Science et l'Hypothèse*), Poincaré showed that the mechanical concept of a constant mass had to be replaced by the idea of mass dependent on velocity and linked to the electromagnetic field (or, at least, acting as if it was related to the field). At the Saint-Louis conference, Poincaré underlined the crisis of Lavoisier's principle. He affirmed that the total electron mass (or apparent mass) was composed of two parts: the mechanical mass and the electromagnetic mass. Poincaré explained that the electron was submitted not only to the mechanical inertia but also to an electromagnetic force, which he later defined as self-induction. In *La fin de la matière*, he clarified:

[...] nous savons que les courants électriques présentent une sorte d'inertie spéciale appelée *self-induction*. Un courant une fois établi tend à se maintenir, et c'est pour cela que quand on veut rompre un courant, en coupant le conducteur qu'il traverse, on voit jaillir une étincelle au point de rupture. Ainsi le courant

---

[13] Poincaré (1901, 142-144).

[14] Cfr. Giannetto, (1998).

[15] For Poincaré, physical concepts have to originate in empirical experience. In this sense it is impossible to consider, e.g., Absolute Motion because it is impossible to determine it in an experimental way. Nevertheless the operational origin of physical concepts in Poincaré's epistemology cannot be interpreted as a form of anti-realism (on the objection to the anti-realism interpretations of Poincaré's epistemology see the conclusion).

tend à conserver son intensité de même qu'un corps en mouvement tend à conserver sa vitesse.[16]

Thus, there are two different reasons which incite the resistance of electrons towards any possible alteration of velocity: its mechanical inertia and its self-induction. The latter is derived from the fact that any kind of alteration of velocity corresponded to an alteration of current. The electromagnetic mass is dependent on velocity and direction, hence it is not constant. In addition, Poincaré emphasized that Kaufmann's experiments showed the inexistence of mechanical mass. Rather, these experiments revealed rather the existence of only electromagnetic mass which is dependent on the electromagnetic field. Moreover, in his 1904 lecture, he claimed that even if Kaufmann's experiments were not confirmed, it would be necessary in any case to consider the mass as variable. Lorentz was obliged to suppose that, in a uniform translated medium, every force was reduced by the same proportion independently of its origin. He did so to preserve the principle of relativity as well as the "indubitable" results of Michelson's experiment. Such a reduction did not only deal with "real" forces but also with the force of inertia: the masses of every particle would be influenced by a translation, behaving in the same way as electromagnetic masses of electrons. Thus, mechanical masses, even if they existed, should be constant. So, the notion of mechanical Mass lost its meaning of basic concept and it was redefined by Poincaré in an electromagnetic manner. Even if Poincaré did not exclude the possibility of conceiving of a mechanical mass, he recognized that mass, like electromagnetic self-induction, was dependent on velocity. Hence, the mass was deprived of its mechanical characteristics.

An analogue treatment was reserved by Poincaré to Ether's notion. His use of this term was often interpreted as a symptom of a classical idea of science and as an epistemological impediment for elaborating a real relativity theory. On the contrary, the term "Ether" was in fact deprived by him of any previous meaning. Since 1899, Poincaré referred to it as a "metaphysical hypothesis" destined to disappear:

> Peu nous importe que l'éther existe réellement, c'est l'affaire des métaphysiciens ; l'essentiel pour nous c'est que tout se passe comme s'il existait et que cette hypothèse est commode pour l'explication des phénomènes. Après tout, avons-nous d'autre raison de croire à l'existence des objets matériels. Ce n'est là aussi qu'une hypothèse commode ; seulement elle ne cessera jamais de l'être, tandis qu'un jour viendra sans doute ou l'éther sera rejeté comme inutile.

---

[16] Poincaré (1902, 246), author's italics.

Moreover, when he describes its physical proprieties he said:

L'expérience a révélé une foule de faits qui peuvent se résumer dans la formule suivante: il est impossible de rendre manifeste le mouvement absolu de la matière, ou mieux le mouvement relatif de la matière pondérable par rapport à l'éther; tout ce qu'on peut mettre en évidence, c'est le mouvement de la matière pondérable par rapport à la matière pondérable. [17]

Only a "matière ponderable" could represent a reference frame for Poincaré. Ether is not considered a material substratum to which phenomena had to be referred. This aspect is confirmed by the fact that in his two papers on electron dynamics, Poincaré uses the term 'Ether' only in the introduction and in order to explain the impossibility of measuring the motion of matter with respect to Ether. In the other parts of these papers there is no reference to Ether and it has no role in the development of either the calculus or the reasoning[18].

5 The importance of experience and experimental data

Through an examination of Poincare's criticism of Mechanics it is possible to understand the importance he gave to experience and experimental data. His reflections frequently arose from experiments (e.g. the experiment of Michelson-Morley and the works of Kaufmann and Abraham). Poincaré often considered the possibility of experimental confirmation to be decisive. Several times, in his scientific works, Poincaré considered experiments capable of condemning the scientific principles and essential to identify the correct theory among a moltitude of possibilities. This aspect is not in contradiction with what he affirmed in his epistemological texts. The role of physical principles and conventions, usually compared to that of geometrical conventions, appeared in Poincaré's work as very complex. As mentioned at the Saint-Louis conference Poincaré spoke about a "crisis of principles". The use of new measuring instruments allowed new experiments and measurement to be carried out which led to results and to conditions of experience that were incompatible with the previous data. Two statement went hand in hand: the generalization of principles involved conventional elements, and, it was necessary to abandon old principles. For Poincaré, there were contexts in which the introduction of *ad hoc* hypothesis was not sufficient to save the principles. Even if they

---

[17] Poincaré (1895).
[18] See: Provost, Bracco (2005).

239

were not directly falsified by experience, they lost their meaning: the experimental proofs attributed only a formal value to them. Thus, the principles did not express anything about physical phenomena anymore. Even if they were not "falsified", they became useless and meaningless.

The experiments acquired a fundamental role in Poincaré's epistemology as starting points for operational definitions. They became the basis upon which it was possible to found a theory that also took into account instruments of measurements, namely a theory about the conditions of knowledge[19].

The undervaluation of empirical data and the misunderstanding of Poincaré's statements about the possibility of constructing different theoretical frames often led to study Poincaré's epistemology through the interpretational categories of his geometrical works. On the contrary, Poincaré's new dynamics originated in experiments and in a criticism of mechanistic perspective. Nature could only be understood through measuring instruments, which were, for Poincaré, indissolubly linked to the assumption of a specific theory on the structure of the world. Human knowledge was impossible without such instruments and was related to the particular world picture on which the theories were founded. Hence, the awareness that there were no dynamics without a physical world view was the real basis of Poincaré's epistemology in physics.

6 Conclusion

Such a relevant and original epistemology cannot be reduced to its main interpretations. About the readings which consider Poincaré as an empiricist, it is sufficient to mention that he continually and explicitly criticized empirical positions. The inadequacy of such interpretations is stressed by the fact that Reichenbach, far from seeing Poincaré as one of his forerunners, criticized him for assigning a sort of «subjective arbitrariness»[20] to conventions. The charges of "antirealism" related to nominalist readings are not justified. For Poincaré, Geometry was nothing but linguistic convention, namely a convenient language among others. His view did not involve an antirealism, but only a rejection of geometrical realism. Even if Geometry indicated a physical reality, for Poincaré it did not coincide with such a reality. Poincaré's statements about the presence of conventional elements in principles could not be seen as "antirealist". Such an affirmation did not conceal a denial of reality, but rather the awareness of

---

[19] See: Giannetto (1998, 180).
[20] Reichenbach (1928).

the limits of theories and principles. The experience could not determine them with certainty; consequently they could be true only within certain limits. Laws and principles were nothing but mathematical forms through which it was possible to describe the world. They were contingent and they changed with the shift of theories in the history of science. Such an evolution, even though it revealed the impossibility of a sure and total knowledge of phenomena, allowed reality to gradually show itself.

Even Giedymin's interpretation[21] cannot be considered completely satisfactory even though it is the most accurate and faithful to Poincaré's texts. In order to direct Poincaré's physical epistemology to the geometrical one, Giedymin found the rise of physical "generalised conventionalism" of Poincaré in the work of Hamilton and Hertz. Thus, he reduced Poincaré's physical thought to what he defined a "Pluralist Conception of Theories". In his opinion, the base of Poincaré's whole epistemology was constituted by a rejection of uniqueness according to which experimental data could lead to different possible theoretical constructions. So, with the aim of founding the totality of Poincaré's thought in his geometrical conventionalism, Giedymin focused his attention on Poincaré's works on geometry and mathematical physics, ignoring Poincaré's physical papers.

In a subsequent paper, Donald Gillies[22] found contradictions in Poincaré's work. In particular, he maintained that Poincaré's scientific practice contradicts his epistemological methodology. Even though Poincaré made a revolutionary advance in his 1905 and 1906 papers[23], such an innovatory step was not followed by his methodological views. On the contrary, in Gillies opinion, such an advance was only made possible by the fact that Poincaré ignored and clashed with his own conservative methodology.

Poincaré's scientific and epistemological activities were never separated. While his scientific works showed the results of an *in fieri* science, his epistemological writings represented a philosophical analysis of classical science. This does not imply the presence of a contradiction between his physical and mathematical researches and his philosophical and popular works. While we must consider Poincaré's writings as a cohesive whole, we should not try to impose one part of Poincaré's though to the entirety of his philosophy.

Poincaré never had the intention of systematically exposing his philosophy. Therefore it makes no sense to look for such an exposition in his writings or to realize an a posteriori synthesis of Poincaré's thought. In

---

[21] See: Giedymin (1982, 1991, 1992).
[22] Gillies (1996).
[23] Poincaré (1905, 1906).

order to understand his thought it is necessary to avoid any kind of synthesis and, on the contrary, to try to understand all its aspects in the context in which they were formulated.

Poincaré's criticism of Mechanism, the operational origin of physical concepts and theories and the importance he gave to experience and experimental data in his physical works, show insubstantiality of the bachelardian concept of 'epistemological obstacle' with regards to Poincaré's works and the necessity of a revaluation of Poincaré's epistemological perspective and of his contribution in special relativity.

References

Born, M. (1964): *Einstein's Theory of Relativity*. New York, Dover Publications.

Einstein, A.; Born, M.; Born, H., (1969), *Briefwechsel 1916-1955*. München, Nymphenburger.

Giannetto, E. (1998): "The Rise of Special Relativity: Henri Poincaré's Works Before Einstein". *Atti del XVIII congresso di storia della fisica e dell'astronomia*, pp. 171-207.

Giedymin, J. (1982): *Science and Convention*, Oxford, Pergamon press.

Giedymin, J. (1991): "Geometrical and Physical Conventionalism of Henri Poincaré in Epistemological Formulation". *Studies in History and Philosophy of Science*, 22, pp. 1-22.

Giedymin, J. (1992): "Conventionalism, the Pluralist Conception of Theories and the Nature of Interpretation". *Studies in History and Philosophy of Science*, 23, pp. 423-443.

Gillies, D. (1996): "Poincaré: Conservative Methodologist But Revolutionary Scientist". *Philosophia Scientiae*, I, pp. 60-69.

Holton, G. (1988): *Thematic Origins of Scientific Thought: Kepler to Einstein*. Cambridge, MA, Harvard University Press.

Miller, A.I. (1973) : "A Study of Henry Poincaré's *Sur la dynamique de l'électron*". *Archives for History of Exact Sciences*, 10, pp. 207-328

Miller, A.I. (1981): *Albert Einstein's Special Theory of Relativity: Emergence (1905) and Early Interpretation (1905-1911)*. Reading, MA Addison-Wesley.

Miller, A.I. (1996): "Why Did Poincaré Not Formulate Special Relativity in 1905?", in H. Poincaré, *Science et philosophie*. Congrès International, Nancy, France 1994, i J.L. Greffe, G. Heinzmann, K. Lorenz, (eds.). Berlin-Paris, Akademie Verlag-Albert Blanchard, pp. 69-100.

Pais, A. (1982): *Subtle is the Lord: The Science and the Life of Albert Einstein*. New York, Oxford University Press.

Pauli, W. (1921): "Relativitätstheorie", in *Encyclopädie der mathematischen Wissenschaften*, vol. V, 19. Leipzig, Teubner.

Poincaré, H. (1880) : "Note sur les principes de la mécanique dans Descartes et dans Leibnitz", par Henri Poincaré, ingénieur des mines, chargé de cours à la Faculté des sciences de Caen, in Boutroux, E. (ed.), *La monadologie*, Edition annotée et précédée d'une exposition du système de Leibnitz. Paris, Delagrave.

Poincaré, H. (1895) : "A propos de la Théorie de M. Larmor". *Eclairage Electrique*, t. III and IV; edited also in *Œuvres*, IX, pp. 369-426.

Poincaré, H. (1898): "La mesure du temps". *Revue de métaphysique et de morale*, VI, pp. 1-13.

Poincaré, H. (1900a) : "Les relations entre la physique expérimentale et la physique mathématique". *Rapports du Congrès international de physique*, tome I, Paris, pp. 1-29; also edited in *Revue générale des sciences pures et appliquées*, 11, pp. 1163-1175; and in *Revue scientifique*, 14, 4ème série, pp. 705-715.

Poincaré, H. (1900b) : "La théorie de Lorentz et le principe de réaction". *Archives néerlandaises des sciences exactes et naturelles*, 5, pp. 252- 278.

Poincaré, H. (1901): "Sur les principes de la mécanique". *Bibliothèque du Congrès international de philosophie*, tome III, Paris, pp. 457-494; also edited in Id., *La science et l'hypothèse*, chap. VI e VII, pp. 142-187.

Poincaré, H. (1902) : *La science et l'hypothèse*. Paris, Flammarion, 2nd enlarged edition: 1907.

Poincaré, H. (1905) : "Sur la dynamique de l'électron". *Comptes rendus de l'Académie des Sciences de Paris*, 140, pp. 1504-1508.

Poincaré, H. (1906) : "Sur la dynamique de l'électron". *Rendiconti del Circolo Matematico di Palermo*, 21, pp. 129-175.

Provost, J.P. ; Bracco, C. (2005) : "Poincaré et l'éther relativiste". *Bulletin de l'Union des Professeurs de Spéciales*, 211, pp. 11-36.

Reichenbach, H. (1928): *Philosophie der Raum-Zeit-Lehre*. Berlin & Leipzig, De Gruyter; also edited in *Gesammelte Werke* Bd. 2. Vieweg Verlag, Braunschweig 1977; english translation: Reichenbach, M., *Philosophy of Space and Time*, (1928). New York, Dover Pubblcation, 1958.

Tonnelat, M.A. (1971): *Histoire du principe de relativité*. Paris, Flammarion.

Whittaker, E.T. (1951-1953): *A History of the Theories of Aether and Electricity*, 2nd edition, vol. 2: *The Modern Theories 1900-1926*. London, Nelson.

Time Travel and the Thin Red Line

Giuliano Torrengo
University of Torino
giuliano.torrengo@unito.it

1 Introduction

In recent articles, Kristie Miller (2006, 2008) has argued that the open future hypothesis is incompatible with the possibility of time travel. Suppose that at an instant *t* a person P pops up out of thin air and sincerely describes herself as a time traveller from the future (suppose, for simplicity, that P is a Wellsian time traveler; see Earman 1995). If at *t* it is nomologically undetermined what contingent facts will take place, it is also indeterminate whether P has actually travelled backward in time or not, because the appeareance of P could be a unexplained phenomenon unrelated to the future. Assuming that backward time travel is a instance of backward causation, the arrival of P would be a determined effect with a undetermined cause. But this is weird, since it violates the plausible "Cartesian" principle that the nature of the cause is always at least as determined as the nature of the effect. The incompatibility between time travel and the open future, thus, is metaphysical rather than logical or conceptual. I sum up Miller's main argument as follows:

1   Backward time travel is a instance of backward causation: the event *e2* of the departure of a person P to the past is the cause of the event *e1* of thearrival of P, and *e1* <*e2*

2   The future is open, and thus it is not determined at *e1*, whether P is a time traveler, namely whether *e2* will take place or some other events whose occurrence is incompatible with the occurrence of *e2*

3   The nature of the cause is always at least as determined as the nature of the effect: if it is determined that an event *e* occurs, then it is determined whether its cause c occurs.

---

4   Either time travel is impossible (at least in the sense that backward time travel never takes place) or the future is not open.

Since the possibility of time travel seems to be a lively option in the present debate in physics, Miller argues that it is the idea of the open future that has to go. Although I take Miller's conclusion to be on the right track, her argument seems to rest on a certain interpretation of the open future thesis, which is questionable. In this note, I will spell out an alternative construal of the open future hypothesis that is compatible with the possibility of time travel.

2 Two Preliminary Remarks

My main point will be that (2) is false, because it can be determined in the present what future contingent events will take place, even if the future is open. Before I get to my main argument, though, I want to defend assumption (1) and (3) against two possible misunderstandings. The first concerns (1) and the notion of backward causation involved in it. There is a sense of "backward causation" that it is incompatible with Relativity theory. According to such a *local* sense of backward causation, the space-time path that connect the cause and the effect is such that it passes through the light cone of the cause and it reaches events in its past. Although this seems to be the sense of backward causation that Miller has in mind, it is not required for argument to take off the ground. Therefore, in order to accept (1), we do not have to make a suspicious physical assumption. Only a *global* sense of backward causation is involved in (1).Backward time travel does not require that a space-time path passes through a future light cone, namely superluminal velocity. What is required for backward time travels is that space-time be a manyfold such that closed time-like curves (CTC) in it are allowed. And, as Gödel has demonstrated for the first time, CTC are allowed by Einstein Field Equations of General Relativity[1].

The second objections concerns (3). One might claim that it is intuitively true that if it is determined that something *is occurring* or *has occurred*, then it is also determined that its cause *occurred*. However, this intuition is strictly linked to the usual order of cause and effect. Once we assume that the temporal order of cause and effect may be inverted, we loose our grip on a principle such as (3). Suppose we try to frame the principle as follows: if it is determined that something *is occurring* or *has occurred*, then it is also determined that its cause either *occurred*, *is occurring* or *will occur*. Is the principle still plausible? One may argue that if the cause of a present event lies in the future, our intuitions about its determinateness are up for grabs. However, this objection is beside the point. It is probably true that cases of backward causations challenge our ordinary intuitions, namely our "tense-bound mentality"[2], but that does not impinge on (3), which claims that we cannot make sense of the *relation* of causality when a cause fails to be determined if its effect is.

3 Two Kinds of Open Future

Let us now focus on (2), and the claim that the hypothesis that the future is open entails that at least some future events are undetermined, in a sense that undermines the possibility of time-travel. What I will call the "minimal conception" of the open future can be spelled out as the conjunction of two intuitive claims:

5   With respect of many aspects of reality, there is *more than one* future alternative

6   All future alternatives are metaphysically on a par.

---

[1] See Gödel (1949). In Gödel's case, CTC are allowed by the effects of the global global rotation of matter in space-time. In contemporary cases, CTC are usually allowed by the topological features of non-simply connected space-time. Both cases would need a more complex formulation of mine (and Miller's) arguments, but nothing hinges on such a simplification. Note that even if backward causation in this global sense is compatible with general relativity, it is a open issue whether there are quantum effects that prevent time travel from occurring. For a discussion of that problem and further bibliographical references see Earman and Wuthrich (2004).

[2] Quine (1987, 197).

These intuitions have been formalised through non-bivalent logics (for contingent future tensed propositions), and semantics based on the branching model of time[3]. I will argue that it is because (i) is usually construed as implying failure of bivalence that the open future hypothesis is claimed to be incompatible with time travel, and that the intuition underlying (i) may be maintained even if we retain a bivalent logic for future contingents. My points will not only concerns terminological issues, since my central argument (§7) is that a bivalent construal of the open future is fully justified even within the metaphysical perspective that is usually assumed by those who defend non-bivalent construals of the open future hypothesis. (Thus, the argument is *ad hominem*, and I grant that it has all the limitations of such kind of arguments — for instance, it does not hold if we drop certain central assumptions).

Firstly, note that the minimal conception of the open future is compatible with many different stances towards the *ontology* of time, such as presentism (the thesis that only present entities exist), the growing-block view (the thesis that only past and present entities exist) and eternalism (the thesis that present, past and future entities exist). What it constraints, rather, is the *topology* of time. The minimal open future view requires that the topology of time is "tree-like", with one single trunk of past events, and many future "branches". Such branches represent metaphysically possible alternatives, and are all temporally connected to the present (they are in the same world), but are not temporally connected to each other. Thus, the minimal conception is incompatible with the linear conception of time[4]. The only ontological constraint, required by (ii), concerns the non-disparity of ontological status of the branches. In other terms, the minimal open future view is incompatible with branching conceptions that make a ontological distinction between the branches. For instance, both a presentist who maintains that *none* of the future alternatives exists (there are only abstract representations of possible futures – they exis, but what they represent does not), and a eternalist who maintains that *all* alternatives exists, hold a position that is compatible with (i) and (ii); but a eternalist who maintains that only one of the branches exist, while there are only representations of its merely possible alternatives, parts company with the minimal conception of the open future.

---

[3] See Belnap et al (2001), and macfarlane (2003), (2008).

[4] Maybe the tenet that the future is open is compatible with "divergence" in Lewis's sense (1986); see Iacona (2007). In that case the claim that the linearity of time is incompatible with the open future hypothesis needs qualification. In any case, if the claim that the future is open is compatible with divergence, it follows that it is compatible with bivalence.

Now, does the minimal conception compel us to endorse a non-bivalent logic for future contingents? No, since there are at least two views that are compatible with it: the standard open future view, which is non-bivalent, and the thin red line view, which is bivalent. We can spell out those two views as follows. The standard interpretation of the open future, which I call the Branching Theory of Time (BTT), adds to the two constraint from the minimal intuition (i.e. A and B below), the tenet that future contingents claims are undetermined (i.e. C below). Whereas the thin red line theory (TRLT) add to the constraints the tenet that there is a branch which is labeled as the "thin red line", and it has a certain role within the semantic such that future tensed claims turns out to be bivalent (i.e. D below)

A. Time has a tree-like topology (Branching Model of Time)
B. All branches have the same ontological status
C. Contingent future tensed claims are neither true nor false
D. One of the future branches is labeled "the thin red line"

$$BTT = A + B + C$$
$$TRLT = A + B + D$$

The difference between BTT and TRLT is not in the metaphysical model of the space-time manifold that underlies them. Both assume the branching model of time[5]. What differentiate the two theories is how they treat the evaluation of future tensed claims about contingent event — such as the event of someone deciding to enter into a time machine or not. Look at fig. 1 and imagine that we are at moment t and we experience event $e0$ of the arrival of person P as if out of thin air in front of us. Suppose that event $e1$ is P entering in the time machine to show up at $e0$, while events $e2$ and $e3$ are events incompatible with P entering in the machine (e.g. P dies or changes her mind) and that on neither branch P subsequently will use the time machine.

---

[5] For simplicity, as it is usual in discussing such issues, I will assume an eternalist ontology, and thus I construe (ii) as implying that all future branches and future events on each branch exist.

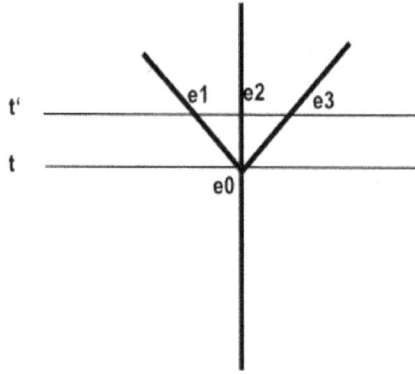

Figure 1

Suppose, furthermore, that at t I claim "P will use the time machine to arrive in front of me at t'". According to BTT, since I am making a claim about some contingent fact in the future, my claim is neither true nor false. And that is so because we find different futures in different branches: in some of them P enters in the time machine, in others she does not. According to TRLT, even if what I claim at t concerns some contingent event, the truth-value of my claim is determined already at t. And this is so, because even if there are different futures for P on different branches, what is relevant for the evaluation of claims about the future is what is to be found on the branch labeled "the thin red line".Let us say that the thin red line is the branch passing form $e0$ to $e1$, then at t it is true that P will use the time machine.

4 Miller's Argument Reconsidered

Note, again, that there are neither differences in the topology of the underlying model (since in both cases the future branches), nor in the ontology (since in both cases the branches are all on a par). The property of *being the thin red line* has no ontological or metaphysical import, it is simply the property of playing a certain role in a *semantics*. Therefore, the thin red line theory is compatible with the minimal conception of the open future as I have highlighted it.

Miller identifies the open future view with the branching theory of time (as is standard). However, it is not the mere topological structure of the branching model that warrants the thesis that it is not determined, at the present, what truth value future contingent claims have. The thin red line view combines a tree-like topology with the idea that claims about the future have a determined truth value even if there are future alternatives. A consequence of that feature of the model, is that it allow us to block Miller's argument against the open future, without giving up the idea that time travel is possible. Consider again P who shows up in front of us as out of thin air at *t*, and claims to have memories of being a time traveler from the future. The intuition here is that if one of the future branches in which P gets in a time machine and travels back in time turns out to be how things will actually go, then P *is* a time traveler, otherwise she is not — maybe her memories deceive her, or she is a cheat. And this intuition is in line with principle (3) that a causal relation with a determined effect requires a determined cause. However, according to the branching theory of time, while it is determined that the arrival takes place at the time of its occurrence *t*, at *t* it is not determined whether its cause (the departure) will take place or not. This situation makes the hypothesis of time travel incompatible with assumption (3) about causality.

I agree with Miller that if we wish to save both time travel and that plausible condition on causality the branching time theory has to go. However, the branching time model has not to follow that fate. And if I am right in claiming that the thin red line theory is compatible with at least a minimal conception of the open future (branching model plus the ontological equality of all branches), then Miller's argument cannot be marshaled against the open future view in general. In the red line model, it is determined at *t* whether the cause will actually take place, and thus whether P is really a time traveller or not, even if time branches toward the future and there is no ontological difference between the branches. Therefore, we are not compelled to choose between an intuitively plausible condition on causality and time travel to save the idea that the future is open.

5 Truth, Determination, Necessity

It is important to notice that in TRLT, it can be determined at *t* that event *e* will take place at a later time *t'*, even if *e* is *not* on every future branch (at *t'*), and thus if it is *not* a metaphysical necessity. Contrariwise, in BTT, the only sense in which a claim about the future can be determined is that is metaphysically determinate, namely it is on every future branch.

To fix terminology, I say that at a time *t* is *determinate* that it will *p* if and only if on every future branch it is true that *p*. Determination in this sense is a kind of metaphysical necessity. Note that it is very implausible to require that a cause of a contingent event is determinate in that sense.

Suppose that the arrival of P at *t* is a contingent matter, and it is true that she is an actual time traveler and not an impostor. It would be unjustified to require that her departure is a necessary event, i.e. that there is no future alternative in which P does not enter in the machine. After all, her arrival is a contingent matter, and thus there are possible words in which P does not show up at *t*. What is required, if P does arrive from the future, is that P *will* enter in the machine. In other terms it has to be *determined* that P will enter, in the sense that it cannot be neither true nor false. It follows that it is determined (as opposed to determinate) at a time *t* that it will *p* if and only if it is true that *p* on the thin red line. Determination in this sense isa mere consequence of how the story turns out to be. (If you like, it is a form of historical necessity, but I think that talking of "necessity" in this case is misleading, and I will avoid it[6]).

Within the TRLT, we can equate truth *simpliciter* with determination in that sense (i.e. being determined). Within BTT, such a notion can only be defined relatively to a branch, since there is no thin red line to play the role of providing a determined truth value to every future tensed claim. It follows that truth *simpliciter* can at best be defined in terms of metaphysical necessity. Indeed, in branching time theory there is no room for a notion of true *simpliciter* or determination that does not collapse on metaphysical necessity (truth on every branch). By contraposition, any future claim that it is not true on every branch, it is not determined (i.e. not true *simpliciter*). That makes the theory incompatible with instances of backward causation that respect (3). Unless the contingent arrival of the time traveler at *t* is a consequence of a necessary cause (but that would be very weird, and in any case, why should always be that way?), the theory will not respect the idea that a cause has to be at least as determined as its effect.[7]

From that difference in the two theory follows also that, according to the TRLT, if it is determined at *t* that an event *e* will happen, then it does *not* have to be a metaphysical necessity that *e* will happen. If determination were tantamount to metaphysical necessity, since the theory implies bivalence for future tensed claims, everything in the future would happen by metaphysical necessity. Any theory that imply that the future is metaphysically necessary in that sense, it is not compatible with the minimal conception of the open future (and thus a fortiori with the hypothesis of the open future tout court), because it clashes with the idea (i) that there are more than one future alternatives. But the thin red line theory does not have any such consequences.

---

[6] See Dorato 1995.
[7] But see Martinez (2011) for a defense of such a idea.

6 Nomological Indeterminism

It may be though that even if TRLT does not imply that the future is close in the sense that it is metaphysically necessary, it is still incompatible with nomological indeterminism, the idea that the laws of nature do not settle everything about the future in advance. But it is easy to see why that is not the case. Causal determinism is the tenet that for every t, what will be the case after t, it is implied by what has been the case up to t together with the physical laws. If determinism is true, then every future tensed claim that is true at a time t is a consequence of what happened before t together with the physical laws. Now, in TRLT, a future contingent claim such as "it will the case that P enters in the time machine" is true (or false) *simpliciter* at the present time t if and only if it is true on the thin red line at some point after t that Penters in the time machine. But nothing in the theory compels us to construe the thin red line as composed by the events that are implied by the past plus the physical laws. Suppose that on the thin red line there is the event that P enters in the time machine, that fact does *not* have to be a consequence of what has happened before t plus the physical laws. Therefore, TRLT is compatible with the negation of determinism, and hence with indeterminism[8].

TRLT is also compatible with determinism, since we can interpret the formalism of the theory in a way that the future contingents that are true *simpliciter* turns out to be those that are determined by laws of physics. But nothing in the formalism as such requires us to assume that nomological determinism is true and to "tune" the thin red line on what is physically necessary. As we have interpreted the formalism of TRLT (and as it is usually presented, even by its detractor), on the thin red line we find what, *as a matter of fact*, will happen. Thus, the theory is perfectly compatible with nomological indeterminism, and it does not compel us to claim that the future is close in the sense that there are no physically possible alternatives to what will happen.

7 Static Models, Dynamic Models

Even if in TRLT there are ways to distinguish being true *simpliciter* from being metaphysically or physically necessary, one may still think that the kind of determination required by bivalence is too strong for allowing future alternatives in a interesting sense. That is why TRLT is often

---

[8] Of course, it may be claimed that indeterminism requires more than the negation of determinism. What is it? If it is negation of bivalence, then my argument fails — but see my point in §7.

criticised as non respecting the idea that the future is open[9]. More precisely, one of the main objection to TRLT can be framed as a dilemma. On the one hand, one may argue as follows: if we can settle at a time t the truth value of claims about later times t' then it seems that the present can "tell" us everything about the future, which means that the future is already settled and closed, that we cannot escape it. In other terms, the thin red line has indeed a "thick" metaphysical ground: the branch that carries the "thin red line" label posseses a certain feature that singles it out already at present. Therefore, it is not compatible with the minimal conception of the open future, in particular with claim (ii) that all future alternatives are metaphysically on a par.

On the other hand, if the thin red line theorist insists that no metaphysical necessity, no physical necessity, no other kind of objective property singles out the thin red line, then the presence of a branch labeled "the thin red line" in the theory is ad hoc. That we can tell already in the present what will be on the thin red line is a brute, ungrounded fact that is left with no explanation. And a theory with such a big explanatory deficit with respect to its main rival is to be abandoned (other things being equal). The question, then, is: what makes *one* branch, rather than another, the thin red line? If it is not possible to answer this question without either begging the question or failing to meet (ii), then we have to dismiss the idea that TRLT can catch the intuition that the future is open, and it is a plausible alternative to BTT.

In order to see why on the one hand there is nothing mysterious in the way the thin red line is individuated and on the other hand the presence of a thin red line in the model is compatible with the constrain on the ontological and metaphysical status of the branches required by the minimal openness intuition, consider the difference between static and dynamic models of time. In the dynamic models, the tensed notion of past, present, and future are taken to be fundamental and not eliminable in terms of indexical notion together with temporal relations. In such models it is a non-relative (to a time) fact that certain events are present, and thus the movement of the present (the fact that what is present keeps on changing) is a genuine feature of the model[10]. In the static models, tensed notions are not absolute: the notion of being present *simpliciter* is undefined, and events can be past, present or future *only* relative to a time (or a event). Thus, tenseless temporal relations (such as before/after and simultaneity) are the fundamental temporal notions. When we talk about the model in tensed

---

[9] Many critiques to TRLT can be read, for instance, in Belnap& Green (1994), and macfarlane (2008). I do not fink any of them knock-down, as I argue in Borghini and Torrengo *forthcoming*.

[10] For a dynamic model of branching time see McCall (1994).

terms, by claiming that a certain event or instant is present, and there are many future branches with respect to it, we are informally and *onlyaccidentally* taking a "internal" point of view to describe certain *tenseless relations* between events. No event in the model is present*simpliciter*and no description of the temporal relations between the events from one internal perspective is metaphysically privileged over any other.

Now, standard branching theories of time embrace tenseless static models[11].Thus, such theories aim on the one hand at dismissing models with a thin red line on the ground that the openness of the future require failure of bivalence for future contingents, and on the other to stick with the (scientifically mainstream) static view of the space-time manifold. Here is the *ad hominem* argument that I have announced in §3: given that the temporal model they assume is static, and not dynamic, the branching time theorist has no reason to claim that the thin red line view is either ad hoc or not so thin as it should. In a dynamic model of time the presence of a thin red line would be, indeed, problematic. When we provide a semantic for future contingents claim on the ground of such models, the present has a advantage over the other times. Therefore, there is an objective difference between the elements of the context of evaluation that belong to the present and those that do not belong to it. Thus, in order to single out a thin red line for the evaluation of future contingent claims with respect to the present, we have to assume that the thin red line has some metaphysically relevant property. The conclusion is that in dynamic models, it is hard to motivate the presence of a thin red line without either breaking constraint (ii) of the metaphysical parity of the branches, or appealing to ungrounded facts.

However, for static models such an argumentation fails. In static models, there is no metaphysical privilege of the present over other times, and there is no objective difference between elements of the context of evaluation that are present and those that are outside of the present. Informally, we can say that the instant of evaluation t is present, and talk about the future and the past; but we should not be led astray from this way of speaking, which — intuitive as it may be — is just a shortcut to talk about the relations between the events in the model. Any "internal" point of view on the model is as good as any other. Therefore, in order to single out a thin red line to evaluate future contingents with respect to a instant t, we are justified in shifting our point of view from the "present" t to its "future". There is no reason to worry about the thin red line having a "think" metaphysical ground, since the fundamental features of our model do not set any distinction between the elements that we find at the moment of

---

[11]See, for instance, Belnap et al. (2001) and Thomasson (1970).

evaluation, and those to be found at other moments. However, the thin red line is not postulated on ad hoc base either. The thin red line is grounded in what (tenselessly) happens, in the events that, as a matter of fact, take places in our world. Indeed, it is not thick (in the sense explained above), but it has a "hard-bottom" metaphysical ground. It is ungrounded only insofar as contingent matters in general are. Why do we find certain events rather than others in it? Well, it just happens that those events, and not others constitute the space-time manifold that the model refers to (*viz.* our universe).

8 Conclusions

If there are no decisive arguments against the construal of the open future hypothesis along the line of the thin red line theory, then time travel and the open future turns out not to be incompatible after all. The core of Miller's argument still stand: any open future conception that drops bivalence for future contingents has either to rule out the possibility of time travel or abandon a very fundamental principle about causality. And since the reasons to hold both alternatives seems quite strong, Miller's argument boils down to a *reductio* of the hypothesis about the future with which we started. However, it is misleading to describe Miller's argument as against the open future. Time travel is not incompatible with *any* conception of the open future, but only with those conceptions that deny bivalence to future contingents[12].

References

Borghini, A.; Torrengo, G. (*forthcoming*): "The Metaphysics of the Thin Red Line" in F. Correia and A. Iacona (eds.), *Around the Tree*. Synthese Library.

Earman, J. (1995): *Bangs, Crunches, Whimpers, and Shrieks. Singularities*

---

[12] Thanks to Andrea Borghini, Baptiste Le Bihan, Manolo Martìnez, Andrea Iacona, Roberto Ciuni, Akiko Frischhut, Sven Rosenkranz, and the anonymous referee for helpful comments and discussions on the ideas behind this paper. Thanks are due also to the organisers and to the participants of the "Open Problems in the Philosophy of Science" Advanced Training School (Cesena, April 15-17, 2010). In particular I wish to thank Vincenzo Fano, Pierluigi Graziani, Valeria Giardino, Elena Casetta and Andrea Sereni for their helpful comments. Thanks also to the partecipants to the *SextoCongreso de la Sociedad Española de Filosofía Analítica*, (Universidad de la Laguna, Tenerife, Spain, 4-6 October 2010) in which a draft of this paper was presented.

and Acausalities in Relativistic Spactimes. Oxford, Oxford University Press.

Earman J.; Wuthrich, C. (2004): "Time Machines". *Stanford Encyclopedia of Philosophy. On-line:* http://plato.stanford.edu/entries/time-machine/

Gödel, K. (1949): "A Remark About the Relationship Between Relativity Theory and Idealistic Philosophy" in P. Schilpp (ed.) *Albert Einstein: Philosopher-Scientist*. La Salle, Open Court, pp. 557-562.

Belnap, N.; Green, M. (1994): "Indeterminism and the Thin Red Line". *Philosophical Perspectives*, 8, pp. 365-88.

Belnap, N. *et al*. (2001): *Facing the Future*. Oxford, Oxford University Press.

Dorato, M. (1995) *Time and Reality: Spacetime Physics and the Objectivity of Temporal Becoming*. Bologna, CLUEB Press.

Iacona, A. (2007): "Future Contingents and Aristotle's Fantasy". *Crìtica. Revista Hispanoamericana de Filosofia*, vol. 39, n. 117, pp. 45-60.

Lewis, D.K. (1986): *On the Plurality of Worlds*. Oxford, Blackwell.

Martínez, M. (2011): "Travelling in Branching Time". *Disputatio*, 4(26), 59-75.

MacFarlane, J. (2003): "Future Contingents and Relative Truth". *Philosophical Quarterly* 53, pp. 321-336.

MacFarlane, J. (2008): "Truth in the Garden of Forking Paths", in M. Garcia-Carpintero and M. Kolbel (eds.), *Relative Truth*. Oxford, Oxford University Press, pp. 81-102.

McCall, S. (1994): *A Model of the Universe*. Oxford, Oxford University Press.

Miller, K. (2006): "Time Travel and the Open Future". *Disputatio* 19(1), pp. 197-206.

Miller, K. (2008): "Backwards Causation, Time and the Open Future". *Metaphysica* 9(2): 173-191.

Quine, W. van O. (1987): *Quiddities*. Cambridge (Mass.), Harvard University Press.

Thomason, R. (1970): "Indeterminist Time and Truth-Value Gaps". *Theoria* 36: pp. 264-281.